LIVERPOOL INSTITUTE OF
HIGHER EDUCATION
THE MARKLAND LIBRARY

An Atlas of Renewable Energy Resources

An Atlas of Renewable Energy Resources
In the United Kingdom and North America

JULIAN E. H. MUSTOE
School of Architecture, North East London Polytechnic

A Wiley–Interscience Publication

JOHN WILEY & SONS
Chichester · New York · Brisbane · Toronto · Singapore

Copyright © 1984 by John Wiley & Sons Ltd.

All rights reserved.

No part of this book may be reproduced by any means, nor transmitted, nor translated into a machine language without the written permission of the publisher.

Library of Congress Cataloging in Publication Data:
Mustoe, Julian E. H.
   An atlas of renewable energy resources.

  'A Wiley–Interscience publication.'
  Bibliography: p.
  Includes index.
   1. Renewable energy sources—Great Britain.
   2. Renewable energy sources—North America.   I. Title.
TJ163.25.G7M87   1984   333.79′11′0941   83-10301
ISBN 0 471 10293 8

British Library Cataloguing in Publication Data:
Mustoe, Julian E. H.
  An atlas of renewable energy resources.
   1. Power resources   2. Renewable energy sources
  I. Title
  333.79   HD 9502.A2

ISBN 0 471 10293 8

Printed in England by Jolly & Barber Ltd, Rugby

Acknowledgements

Many people have helped me with the task of writing this atlas.

I am particularly indebted for information and advice to Mr S. P. Carruthers of the Department of Agriculture and Horticulture at the University of Reading, Mr J. P. Cowley and Mr F. Rawlins of the Meteorological Office, Mr G. Foley of the International Institute for Environment and Development, Dr D. C. Hodges of Hull College of Higher Education, Mr R. B. Tranter of the Centre for Agricultural Strategy at the University of Reading, Mr T. Tung of the Canadian Department of Energy, Mines and Resources, and Dr J. Wheildon of the Imperial College of Science and Technology. A special word of thanks is due to Dr J. A. Manuel of the Department of Applied Physics of the North East London Polytechnic. While scrutinizing my draft text with a critical eye he detected frequent omissions and corrected many mistakes.

For expert mapmaking advice and elegant cartographic draftsmanship I am much indebted to Mr G. E. D. Cole of the North East London Polytechnic and Miss E. L. Orrock of De Havilland College.

But no amount of help can entirely preserve an author from making mistakes, and I must lay exclusive claim to any errors of fact or opinion to be found in these pages.

Contents

Chapter 1	Introduction	1
Chapter 2	Solar Energy	15
Chapter 3	Wind Energy	47
Chapter 4	Wave Energy	65
Chapter 5	Ocean Thermal Energy	77
Chapter 6	River Energy	87
Chapter 7	Biofuels	95
Chapter 8	Geothermal Energy	129
Chapter 9	Tidal Energy	145
Chapter 10	Conclusion	163
Appendix 1	Conversion Factors	173
Appendix 2	Thermal Value of Fuels	175
Appendix 3	Multiples	177
Appendix 4	Derivation of Maps	179
Appendix 5	Sources of Figures and Diagrams	181
Appendix 6	United Kingdom Counties	183
Appendix 7	North American States and Provinces	185
Appendix 8	Copyright Acknowledgements	187
	References	189
	Bibliography	195
	Index	197

Chapter One
Introduction

In recent years much has been published about the techniques of harnessing renewable energies. The design and construction of windmills, solar collectors, tidal barrages and turbines, biogas plants, and geothermal power plants have been discussed in great detail and each of these subjects has expanded into a specialism in its own right. A measure of the importance of this work is the fact that many governments support continuing programmes of research into renewable energy technologies. Knowledge and experience is growing rapidly as results from pilot projects, as well as theoretical work, is published and disseminated. Comparatively little attention, however, has so far been devoted to the question of the size of the renewable energy resources which these technologies are designed to exploit. It is the purpose of this atlas to assess how much renewable energy is available to us, and to describe the pattern by which its availability varies from place to place.

The idea of energy is an abstract one, for energy itself cannot be weighed, shaped, or seen. The energy said to be possessed by a hot object, a shaft of sunlight, or by a moving body of water can only be demonstrated by its effects. A hot object will move a thermometer to a point on its scale, the energy of sunlight can be detected by a thermometer or by the thermally-sensitive nerve endings in our skin, while water in motion is seen to be capable of setting a water wheel in rotation. In all these cases the notion of energy is invoked to account for the observed effect. Energy can therefore be regarded as a type of currency, devoid of substance but exchangeable through all the physical processes of the universe.

Energy is usally defined as 'the capacity for doing work'. The thermodynamic unit of measure is the joule, so called in memory of the nineteenth-century English physicist James Joule. When energy manifests itself in such a way as to exert a force it is said to be performing work. The heat energy lost by a body of steam when it does work by, for instance, moving a piston are directly equivalent and the unit of the joule is therefore used to quantify both energy and work. When energy is transformed at the rate of 1 joule per second the rate of energy flow is described as occurring at a power of 1 watt. The inventor of the first effective steam engine, James Watt, is commemorated in this unit of measure.

The joule and the watt are rather small units of measure when processes on an industrial scale are under consideration. Even to boil a kettle about 600,000 joules of energy are required while a litre of petrol when burned liberates roughly 40 million joules. But energy transformations occurring in industrial operations are so large that even the megajoule, equal to 10^6 joules, is too small a unit. Energy considered as a national resource is therefore measured in subsequent chapters in the very large unit of the petajoule of 10^{15} joules. All references to energy in this atlas are made in multiples of the joule in order that the various forms may be compared with one another easily and directly.

Energy is often spoken of as being used or consumed, and this commonly accepted usage will frequently be encountered in the pages that follow. However, it should be realized that in strict parlance energy is never consumed, for it is an indestructible commodity. Correctly understood, energetic processes involve only the conversion or transformation of energy from one form or level to another. The second law of thermodynamics tells us that the result of all energy transactions is to reduce a more highly ordered energy state to one of less order and increased randomness. When a piece of coal, for instance, is burned, the highly-ordered energy of its chemical bonds is not consumed, but rather it is converted into an equal quantity of lower grade and more random heat energy. The sense in which the energy of the piece of coal can be said to be used or consumed is that after it is burned the quantity of energy that remains in a form useful for human purposes is smaller. In the long run, and operating at a cosmic scale, this type of process will eventually reduce all energy to the same level of randomness and so lead to the heat death of the universe.

In the meantime we may content ourselves by observing that renewable energy resources on earth are characterized by the arrival of a supply of energy at a rate sufficient to replace that extracted from them by mankind. The sources of this flow of energy, a part of which we divert and convert for our own use, are the sun, radioactivity within the earth, and the force of gravitation. While it is true that coal, oil, and natural gas are also continuously replenished by the flow of energy from the sun, the processes by which they come into being are so slow that for practical purposes these sources of energy have to be regarded as finite in amount. The distinction commonly made between fossil fuels as an endowment of energy capital and renewable energies as an income of energy is therefore a valid one.

The renewable energy regimes of two areas of the world are described in this atlas. The British Isles lie between 49 and 61 degrees north latitude at the eastern margin of the North Atlantic Ocean. The rock formations of which the islands are constructed are stable and mountain building activity, apart from some vulcanism in Tertiary times, has long been absent. During this extended period of relative geological tranquility the mountains of the British Isles have been eroded down to their present modest heights. Ben Nevis, the highest mountain in Britain, attains to only 1343 m.

The climate of the islands is maritime and is dominated by the proximity of the ocean. No part of the British Isles is more than 150 km from the sea. The North Atlantic drift current, carrying with it warm waters of tropical origin, ameliorates the climate of Britain as it does the whole of northwestern Europe. After so long a journey the waters are well mixed and the steep ocean temperature gradients characteristic of tropical seas have been eliminated. Winters, under the influence of the Atlantic, are mild and summers are cool, while a moderately humid westerly air stream prevails at all times of the year. The ecological climax of all parts of Britain, except the highest mountains, is oak or pine forest.

But though the climate is mild the oceans surrounding the British Isles are relatively turbulent. The circulation of winds in the North Atlantic places the islands at the downwind side of the ocean. The waves which arrive on western coasts are therefore larger than the world average. In southern and western Britain the range of the tides is also higher than the world average. The estuary of the River Severn, where the average rise and fall is as much as 9 m, is the most favourable tidal energy site in Europe and one of the largest in the world.

The second of the two regions of the world discussed in this atlas is the North American continent, composed of the United States and Canada. It is many times more extensive than the British Isles. From the northern part of the Canadian arctic archipelago to the southern tip of Texas is 55.5 degrees of latitude while the distance from the Aleutian Islands to Newfoundland is some 7750 km. The natural conditions, and therefore the renewable energy regimes, of such a large part of the earth's surface exhibit much more variety than is to be found in the relatively compact space of the British Isles.

The eastern half of the North American continent is composed of stable rock formations, but in the west the Rocky Mountains are an active area of geotectonic activity. A chain of volcanoes extends from the Aleutian Islands through western Canada and the United States southward into Mexico. Shallow hot rock in this region is already exploited for electricity generation and this resource makes a small contribution to North American energy supplies. The Rocky Mountains are a young and therefore lofty mountain range. Their highest point, Mount McKinley, reaches an altitude of 6190 m and the Rockies are the source of a number of large and powerful rivers. In the east the Canadian Shield and the Appalachian Mountains also contain large resources of river energy.

The wave and tidal energy resources of the oceans adjoining the North American continent are greatest at higher latitudes. Tidal ranges are very large in southern Alaska and in the Bay of Fundy, but in southern California and the Gulf of Mexico, where tides are small, there is little prospect of obtaining electricity from coastal barrages. Average wave heights are small at low latitudes, but a considerable wave resource exists at middle and high latitudes, and in particular off the western coast of Canada. Ocean thermal gradients provide a large ocean thermal resource off the southeastern region of the United States.

Unlike the British Isles, which are climatically remarkably uniform, North America exhibits every variety of climate and weather. The climates of the east and west coasts of North America are moderated by their proximity to the Atlantic and Pacific oceans but the central parts of the continent have the very hot summers and extremely cold winters characteristic of a continental climate. In southern California Death Valley is one of the driest places on earth while rainfall in the coastal regions of British Columbia equals the wettest tropical regions. Mean temperatures range from minus 10 °C in arctic Canada to 22 °C in southern Arizona. Such wide variations in temperature and humidity produce every kind of natural vegetational cover, from arid desert to grassland, forest, arctic tundra, and, in Florida, tropical swamp. It follows that the availability of renewable energies is markedly different in the various regions of the North American continent and that few broad generalizations about their occurrence can be made. The distribution of the respective energy resources, in both the British Isles and North America, is described in detail in subsequent chapters.

The term *renewable energy* can in its widest interpretation be taken to embrace some dozen sources of energy which are, or can be, replenished by the forces of nature. This atlas is concerned with the eight most significant of these resources, which are:

Solar energy
Wind energy
Wave energy
Ocean thermal energy
Tidal energy
River energy
Biofuels
Geothermal energy

Attention could perhaps be given to other areas of interest such as exploiting ocean salinity gradients, high altitude atmospheric jet streams, or the extra terrestrial solar flux. However, too little is at present known about the pattern of their occurrence in nature, or the techniques for harnessing them, to enable a quantitative analysis to be made. The bounds of this enquiry have therefore been drawn to include only the eight sources of renewable energy whose size and availability is capable of informed assessment.

The size of an energy resource is a function of the methods which are available to exploit it. The techniques needed to extract coal or oil from the earth, for instance, are well understood and it is therefore possible to estimate the size of the resource which is accessible by these means quite accurately. The wind energy resource of the United Kingdom or the amount of biofuels available in North America, to take two renewable examples, must similarly be measured by reference to the state of windmill technology or to the productivity of biogas plants. It is not the purpose of this atlas to investigate the engineering aspects of renewable energy in any detail, but each chapter is furnished with a description of the conversion process upon which the resource assessment is based. In general a somewhat sceptical attitude has been taken as to the performance to be expected from conversion processes in order that the size of the eight renewable energy resources under investigation should not be exaggerated.

For the purposes of economic life an energy source capable of producing high temperatures is highly desirable and in many cases it is essential. This follows from the fact that all thermodynamic processes operate more efficiently when the difference in temperature between the source of energy and the sink to which it is rejected is large. A boiler, for instance, will heat water more quickly when its fuel burns to create a temperature much above the temperature of the water to which its heat is transferred. Similarly, the lower the temperature of the inlet air of a gas turbine the larger will be its output of work. Much technical effort is consequently directed

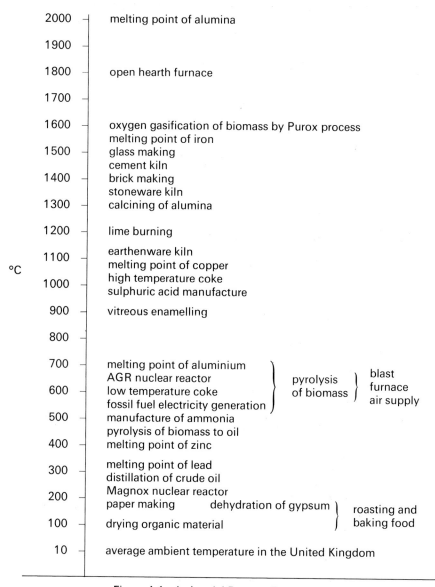

Figure 1.1 Industrial Process Temperatures

to achieving high temperatures, and therefore high efficiencies, in industrial thermodynamic operations. Furthermore, many physical processes which are economically significant will only occur at high temperatures. Smelting ores and firing pottery are two examples. The temperatures needed to perform a number of industrial operations are given in Figure 1.1 and 1.2.

Machines operating at high power rates can be smaller than low power machines capable of the same amount of work. High temperature steam turbines, for instance, are notably compact for their output and gas turbines, whose operating temperatures are at the limit set by material science, are smaller still. Machines operating at a high temperature and power are therefore usually cheaper to construct, and this makes them appear to be more economic than machines of comparable output working at a lower wattage. They are, in fact, more economic provided that the high temperature heat source can itself be obtained cheaply and, at the centre of the present discussion, without exhausting the Earth's large but finite resource of fuel.

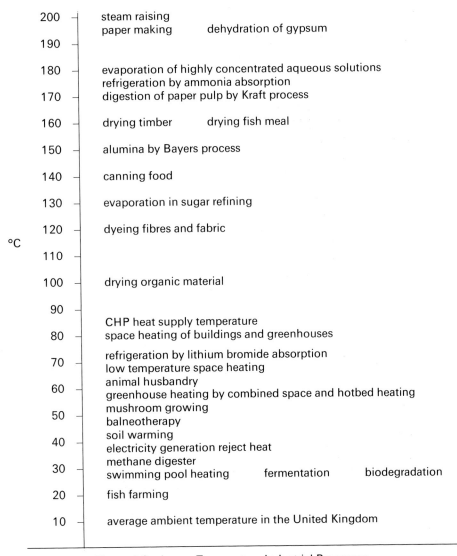

Figure 1.2   Lower Temperature Industrial Processes

6

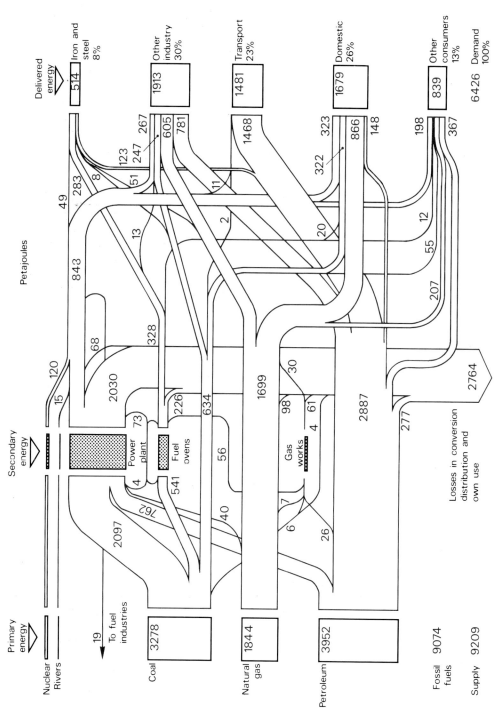

Figure 1.3 Energy Flow in the United Kingdom 1979

At this point one encounters the central dilemma of the modern industrial economy. Fossil fuels when burnt deliver energy at a high temperature. They can reach any part of the temperature scale of Figure 1.1 and the devices needed to utilize them are small and productive. But the stock of the precious fuels by which the world's industries are driven is being rapidly depleted, and the difficulty of extracting what remains is ever increasing. The realization has been forced upon us that the apparent economy of fossil fuel engines is illusory, and is based upon the tacit assumption that the supply of these fuels will last for ever. This is equivalent to saying that they can be regarded as energy income rather than as a capital endowment of energy. The difficulties arising from the depletion of the Earth's stock of fossil fuels, particularly oil and natural gas, are the cause of worldwide concern. It is indeed the most serious problem by which industrial society is beset. Much attention is therefore now being given to a search for additional energy supplies and to developing substitutes for fossil fuels. This atlas is a reflection of the consequent rise of interest in energy resources that, unlike coal, oil, and gas, replenish themselves by the operation of the forces of nature and so do not become depleted with use.

At the time of writing, renewable energy resources contribute but little to the economic life of the United Kingdom. A mere 15 PJ was supplied by river power to the British electricity market in 1979, only 2 per cent of deliveries to electricity consumers in that year. Nearly all was generated on the rivers of the northern and western districts of Scotland. Figure 1.3 represents by means of a Sankey diagram the flow of all energies, including river energy, through the economy of the United Kingdom in 1979.

Energy entering the economy is represented by the quantities assembled at the left-hand side of the diagram. The values shown for the thermal content of coal, gas, and petroleum are net, exclusive of the energy consumed in extracting these fuels, in refining them, and bringing them to market. Movement of energy through the economy is represented by diverging and converging bands, whose widths are proportional to the size of the flow. Energy demand, which in 1979 totalled 6426 PJ, is made up of the five categories at the right of the diagram. Demand is met by the parts of each primary energy resource which reaches the consumer as delivered energy. Beyond the right-hand margin of the diagram a proportion of delivered energy is made use of in the equipment and appliances of consumers. Here it fulfils its purpose of providing heat, light, and motive power for the better conduct of life and work. It is sobering to realize that, at present, this useful energy amounts to only about half of energy delivered. The remaining 50 per cent is dispersed to the environment as waste heat. Heat lost from buildings and by vehicles accounts, at present, for a large part of the delivered energy that is wasted in both the United Kingdom and in North America.

Renewable energies make a proportionately slightly larger contribution to the North American economy than they do in the United Kingdom. In Canada and the United States, because of the much greater size and power of the rivers of that continent, the proportion of electricity demand met from fluvial sources is 22 per cent of a much larger quantity of electricity. The small geothermal electricity generating industry of the western United States, shown in Figure 1.4 to produce 19 PJ a year, makes at present a statistically insignificant additional contribution to North American supplies. The flow of energy through the North American economy is shown in Figure 1.4.

The oft-repeated assertion that renewable energy resources are of only minor significance may therefore be taken to be correct in the circumstances prevailing at the present time. But the existing energy situation, as will be shown in the concluding chapter of this atlas, is not immutable and in fact cannot continue unchanged. Meanwhile it is necessary to consider some aspects of our present-day energy economies in more detail.

The energy movements depicted in Figures 1.3 and 1.4 originate principally from the destruction of the chemical bonds of certain hydrocarbons occurring naturally in the crust of the Earth.

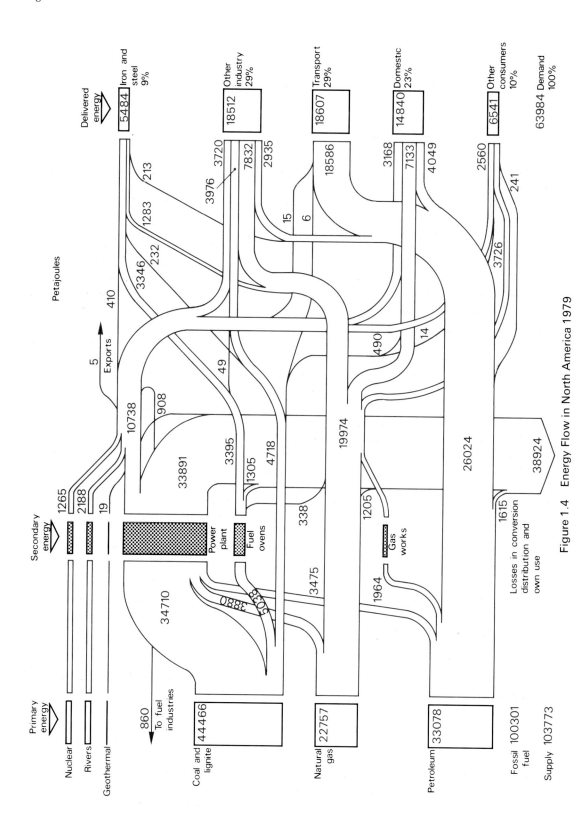

Figure 1.4 Energy Flow in North America 1979

Coal, lignite, natural gas, and crude oil are the fossilized remains of ancient plants and animals, and it is on account of their method of formation that these fuels have been given their generic title. Other parts of the energy flows begin in the kinetic energy of flowing river water, the heat of the inner parts of the Earth, and in the forces existing within the nuclei of uranium and plutonium atoms. The last two energy resources, geothermal and nuclear, are ultimately finite although they are so large that they could hardly be exhausted by any conceivable level of human activity. Forgetting for a moment the obstacle in the way of gaining access to them, nuclear and geothermal energies can best be described as quasi-renewable resources. River energy is, however, a truly renewable resource for a river's water only flows because it is continuously replenished by meteorological processes. Collectively the energy sources at the left of the Sankey diagrams are described as primary energies in recognition of the fact the flows of energy in the world's economies originates with them.

Rather than being delivered directly to consumers, primary energy can, as a matter of convenience and utility, be transformed into other secondary forms of energy that do not occur in nature. Coal can be gasified and distributed as a fuel of medium thermal value. Coke, the nearly pure carbon residue of coal after roasting, is used as a fuel for smelting iron ore and to a small extent as a boiler fuel. But by far the most useful, extravagant, and controversial secondary fuel is electricity. Electricity is from the standpoint of the consumer an almost ideal fuel. It is versatile, clean in use, easily controlled, and can be utilized in simple appliances. Sometimes it seems to be a commodity possessed of almost magical properties. Electricity is everywhere much in demand, and its production forms a large part of the economic activity of all industrialized nations. But electricity generation is a process beset by difficulties and fraught with many disadvantages.

The most conspicuous drawback, evident from Figures 1.3 and 1.4, is the immense waste of energy resulting from electricity generation. The flow of waste heat out of the economy, represented by the flows toward the bottom of the diagrams, is largely made up of electricity power station losses. This is not caused by shortcomings in design and engineering, but follows from the fact that thermodynamic considerations limit the thermal efficiency of electricity generation to a maximum of approximately 38 per cent. The most modern power stations approach this figure closely, although average plant efficiencies in 1979 were 27 per cent in the United Kingdom and 19 per cent in North America. Consequently more than two-thirds of the thermal content of fuel burnt in power stations is rejected to the environment and lost. Further losses in distribution bring the overall efficiency of the British system down to 25 per cent of the primary fuel invested. Thus it is that in the United Kingdom the heat wasted by Central Electricity Generating Board power stations is 64 per cent of the thermal value of all the coal delivered to market by the National Coal Board. An even more fantastic situation prevails in North America where 78 per cent of the value of all coal produced is lost in the course of electricity production and distribution. These enormous losses will continue as long as waste heat from power stations is simply dissipated to the environment. Measures to reduce this waste by putting the rejected heat to use are discussed in the concluding chapter of this atlas.

In use electricity gives an impression of cleanliness. It is invisible and weightless, produces no fumes and leaves no ash or residue. But from a national standpoint it can be seen that electricity, while removing dirt from the surroundings of the consumer, concentrates pollution at the site of production. Many tons of carbon dioxide, sulphur, nitrogen compounds, and dust are discharged into the atmosphere by a power station's boilers in the course of a year's operation. As well as polluting the neighbourhood of the power station the effects of burning coal in large quantities are felt hundreds of miles from the place of origin. Degradation of rivers and forests as far away as, for example, Norway and Sweden can in part be attributed to the output of British chimneys. Additionally, energy rejected by the power production process has to be transferred by heat exchangers to air or water heat sinks whose temperature is thereby artificially raised. Power stations are the main source of thermal pollution of air, rivers, and coastal waters.

Pollution problems in their most acute and intractable form, however, arise from electricity generation by nuclear power. This dangerous technology is increasingly favoured by the electricity generating industries of many countries because uranium and plutonium are thought to be the only fuels in large enough supply to sustain the future of our industrial economies. It will be shown, in the concluding chapter of this atlas, that this assumption is ill founded. In the meantime it must be noted that the fuels and some of the waste products of the nuclear power fuel cycle are chemically toxic, radiotoxic, carcinogenic, mutagenic, and extremely difficult to handle safely. It is universally agreed that these substances must be kept separate from the biosphere for very long periods of time, and great care has always been taken to ensure that their isolation is as permanent and complete as possible.

Since the activity of all radioactive isotopes declines asymptotically with time it can be said, without producing much enlightenment, that they never lose their radioactivity completely. For practical purposes the period of time taken for the activity of an isotope to decline to half its starting level, a parameter known as its half life, is used as a measure of its radioactive longevity. The method of determining the half life of an isotope is illustrated in Figure 1.5. The half lives

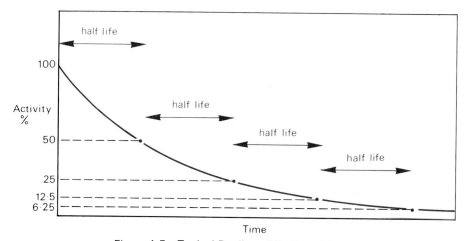

Figure 1.5   Typical Decline of Radioactivity with Time

of uranium–235 and plutonium–239, the two most important components of nuclear fuel, are 710 million and 24 thousand years respectively. Half lives of isotopes appearing as wastes in the nuclear fuel cycle vary from eight days for iodine–131 to 170 million years in the case of iodine–129. Small quantities of uranium and plutonium remain after nuclear fuel is reprocessed and they make their contribution to the radioactivity of the wastes. The progressive decline in the level of activity of the waste produced by a thermal nuclear reactor is shown by the curve of Figure 1.6. It will be observed that it is a slow process occurring over a long period of time.

The biological hazard posed by a radioactive substance is not dependent upon its level of activity alone. Living organisms will absorb and retain one chemical more readily than another. Active isotopes, like strontium–90, which are retained in the body at sites near the bone marrow where the blood is manufactured, are more dangerous to life than, for example, the radioactive but chemically inert gas krypton–85. The risk to life is therefore a function of physiological behaviour as well as radioactivity.

A commonly-used method of establishing comparability between radiotoxins, and between them and other hazardous substances, is to use an index of toxic potential related to the ingestion hazard of the substance. The toxic potential is defined as the volume of water which is needed to dilute a given quantity of the substance to a safe concentration in water to be drunk by the public. Maximum permissible concentrations in water of all radioactive materials are determined and published by the International Commission on Radiological Protection. When this index is employed a substance hazardous to human life is revealed as one requiring a very large quantity of water for its safe dilution. It is possible by reference to Figure 1.6, which uses the index of toxic potential related to ingestion hazard, to assess the biological danger posed by the wastes which result from the operation of the nuclear fuel cycle.

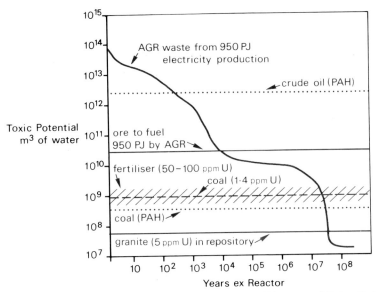

Figure 1.6 Toxic Potential of Nuclear Wastes and Other Substances

The declining curve represents the toxic potential of the waste produced in an advanced gas cooled (AGR) nuclear reactor from the generation of 950 PJ of electricity. The uppermost firm line is drawn across the graph at the toxic potential of the quantity of uranium ore from which the nuclear fuel needed to generate that amount of electricity was obtained. This shows that the activity of the waste will decline to the level of the ore from which it came after about 7000 years. But uranium ore, which does not occur in Britain, is itself a biologically dangerous substance. It is therefore of interest to know how much time must elapse before these wastes are no longer a danger to the British environment.

The most favoured technique for returning radioactive wastes to the environment is by means of glassification followed by burial in a stable rock formation. In what is known as the Harvest process the wastes would be mixed with molten glass and then cooled to form cylinders measuring 2 m in length by 0·5 m in diameter. After being encased in stainless steel the blocks would ultimately be disposed of in an underground repository excavated deep within a geologically quiet formation of granite. All being well they can then be forgotten. But the question arises of how long will it be before these wastes present no more risk than does the granite within which they are entombed.

The lower firm line on Figure 1.6 shows the toxic potential of the volume of granite, some

770,000 m³, which would be occupied in an underground repository by the glassified wastes from the production of 950 PJ of nuclear electricity. The radioactivity of the granite is produced by its content of five parts per million of uranium. The point of intersection of this line with the curve of the decay in the activity of the wastes shows that we must wait 10 million years before they are no more dangerous to life than is the granite rock.

Proponents of nuclear power are sometimes moved to observe that their technology is required to satisfy much more stringent safety standards than are imposed upon other comparable industrial activities. Figure 1.6 shows that this complaint is justified, at least in part. Other industries can be seen from the graph to produce a potential health hazard nearly as great as, and in one case greater than, that resulting from the nuclear fuel cycle. Agricultural fertilizers, for example, are manufactured from hydrocarbons and consequently contain small quantities of uranium and its decay products. The quantity of fertilizers used each year by the British agricultural industry possesses the toxic potential lying within the horizontal hatched band at about 10^9 m³ of water. The biological hazard of posed by this source of uranium ought to be a matter of much concern because the use of artificial fertilizers introduces radioactivity from the uranium directly into the human food chain.

Coal contains a large number of impurities including uranium and a number of poly-aromatic hydrocarbons. The toxic potential of the uranium and its decay products that are contained in the coal needed to generate 950 PJ of electricity in a conventional steam power plant is shown by the dashed line drawn at just above 10^9 m³ of water on Figure 1.6. The chemical toxic potential of this amount of coal, deriving from its content of carcinogenic poly-aromatic hydrocarbons, is shown by the lower dotted line. The upper dotted line corresponds to the PAH index for crude oil and shows that oil produces a significant health risk from cancer. It is true to say, therefore, that other processes are potentially almost as dangerous to public health as is the nuclear power industry. It is long past the time that they were required to operate to the same stringent standards of public health as the nuclear technologies.

A certain care is needed when interpreting the information given in Figure 1.6. In the first place, it should be observed that both axes of the graph are logarithmic. Consequently, differences between values are much greater at the higher ends of the scales. The apparently steep decline in the toxic potential of nuclear wastes after 10 million years, for instance, is due to the fact that the last division on the scale of time occupies 90 million years as against the 10 years represented by the first interval.

It is necessary to remember too that the index of toxicity, against which the various pollutants are measured, is one of potential hazard. The actual hazard which would occur in practice would depend upon the pathway by which the substance reached the public. Radioactivity leaking from an underground repository, for example, would be partly absorbed by the rocks within which the wastes were stored. Furthermore, the rate at which it migrated to reach the biosphere would be affected by movements of groundwater. Both these factors are specific to a particular site, and a complete risk analysis must therefore take into account actual pathways as well as toxic potential.

Figure 1.6 shows the toxic potential of the poly-aromatic hydrocarbons that are contained in coal and crude oil in their unprocessed state. Were they to be burned much of their content of these chemicals would be released to the atmosphere where they would be exposed to the action of sunlight. Since all hydrocarbons are degraded to a greater or less extent by ultraviolet light the actual risk that they pose to life is dependent upon their method of release and upon the prevailing meteorological and geographical conditions. Were sufficient information available it would be possible to draw more detailed diagrams, corresponding to Figure 1.6, to describe the actual risks presented by the various fuels when they are used in particular ways and under specific circumstances. It is unfortunate that full risk analyses of this type cannot, in the present state of knowledge, be carried out. A complete investigation of the pathways followed by all pollutants in the biosphere would be a very laborious task and would take a long time. It is long overdue, however, and it would produce results that are badly needed.

A difficult and hazardous undertaking like nuclear power might perhaps be acceptable were the environmental risks fully understood and were perfectly effective methods of conducting nuclear processes on an industrial scale a practical possibility. Because many substances used in or produced by the nuclear industry remain radiotoxic for so long, releases of them to the environment are, in effect, cumulative. Radioactivity from any plutonium released to the biosphere even far in the future will be added to the amount from this source that is already there. Therefore, a plausible nuclear power industry must be able to maintain the complete reliability of its processes for thousands of years.

But the affairs of the world take place against a background of human error, organizational lapses, ignorance, criminal activity, war, and the tides of history. No human activity can, for these reasons, attain and maintain the perfect standard of performance demanded by nuclear technologies. Every possible effort has been made during the last 40 years to keep nuclear engineering at a high level of performance. But even during the relatively tranquil historical period in which we live, when mankind's skills and social organization have been brought to an unprecedented pitch of development, the difficulties involved with nuclear power have been so great that innumerable leaks of radioactivity have occurred. There have, in addition, been a number of serious accidents and a few near catastrophes. In practice, therefore, the problems involved in operating a nuclear power industry in the real world over long periods of time ought to be recognized as grave and insoluble.

At present nuclear power furnishes only 1·3 per cent of the primary energy delivered to the British economy. In North America the proportion is 1·2 per cent. Both economies are today sustained almost exclusively by fossil fuels. Coal is the most important fuel in North America, followed by petroleum and then natural gas. In Britain slightly more petroleum than coal is consumed with natural gas again lying in third place. In addition to 2097 PJ of coal, electricity generation consumed about 19 per cent of the petroleum entering the British economy in 1979. The proportion in North America was 11 per cent. In both cases the largest consumer of petroleum is the transport sector. Trains, ships, lorries, and cars rely upon petroleum for nearly all their supply of fuel. In North America, as a result of more dispersed settlement patterns, transportation accounts for a larger share of total demand than it does in the United Kingdom. The proportions are 29 per cent and 23 per cent.

Natural gas, a substance possessing almost all the virtues to be sought in a fuel, plays a large part in the North American economy where it furnishes nearly a quarter of all primary energy. In the United Kingdom the proportion is about one-fifth. Reserves of natural gas are, however, limited. In the United States, where the fuel has been used on a large scale for 40 years, supply difficulties are enforcing a decline in its use. Recently discovered gas fields in the North Sea have resulted in the almost complete replacement during the 1960s of manufactured gas by natural gas in the British economy. The medium estimate of North Sea reserves postpones a gap between gas supply and demand in Britain to the turn of the century. Thereafter the rate of use of natural gas in the United Kingdom is also expected to decline. It will be shown in the concluding chapter of this atlas that the exploitation of renewable energies can accommodate the economy to the progressive depletion of natural gas supplies both in Britain and in North America.

In the following chapters the geography of the eight principal renewable energy resources is described and illustrated in map form. The size of each resource is provisionally assessed by comparison with 1979 levels of energy consumption. In this way a rough idea of whether the resource is large or small can be obtained. For an accurate assessment, however, feedback effects must be taken into account. Generating electricity by tidal barrages, for instance, would feed back to the economy by reducing the need for coal deliveries to power stations. In the concluding chapter of the atlas, therefore, all the available and quantifiable renewable energy resources are placed in the context of the economy in which they would function. It is gratifying to be able to report, in advance, that they could make an important and very substantial contribution to the future of our economic life.

Chapter Two
Solar Energy

THE NATURE OF SOLAR ENERGY

From earliest times the Sun has been recognized as the source of life-giving heat upon Earth and as the basis of the continuity of human existence. In the third millennium the Sumerians and later the Akkadians established Utu and Shamash as the sun gods of justice and prophecy. The god Ra, whose symbol was a sun disc, was from the most remote times the chief figure in the pantheon of the bronze age Egyptians while for the classical Greeks Apollo was the god of light, knowledge, reason, and harmony. These beliefs, dating from the dawn of recorded history, are comparatively late manifestations of myths as old as human consciousness, for the Sun has always been an object of intense interest to mankind. It may be said that in recent years some aspects of these ancient sun cults are returning to our attention as the Sun is perceived, literally and scientifically, to be the origin of the energy upon which civilization as well as all earthly life must depend.

Even when deprived by the modern viewpoint of its divine attributes the Sun remains a most remarkable object. Its enormous mass, 333,222 times that of the Earth, is responsible for the gravitational field controlling the motion of the planets that make up the solar system. The Sun's mass is contained in a sphere 696,000 km in radius, some 110 times the radius of the Earth. Four-fifths of its mass consists of hydrogen and nearly all the remainder is helium. The Sun's average density of $1 \cdot 41$ g cm^{-3} is slightly less than one quarter that of the Earth.

The material of the Sun is unevenly distributed through its depth. The internal gravitational force of so large a mass produces by compression at its centre a core whose density is 90 times that of water and where temperatures of 15×10^6 °K are sustained. Surrounding the core, whose radius is 225,000 km, is a shell of hydrogen some 471,000 km thick forming the fuel store for the atomic furnace at work within. It is the outer surface of this mass of hot hydrogen, the photosphere, that forms the visible disc of the Sun. At the photosphere the density of the solar material has fallen to only 10^{-7} g cm^{-3}, one ten-millionth the density of water. The structure of the Sun is maintained in equilibrium against its own gravitation by the balancing force of radiation pressure acting outwards from the core.

In the conditions that exist within the core of the Sun the thermal motion is so violent that atomic structures are unable to survive. Charged sub-atomic particles are packed together in the core into what is known as a plasma. In a sufficiently hot plasma, such as that within the Sun, a nuclear fusion process will occur which can be represented by the equation:

$$4_1^1\text{H} \rightarrow {}_2^4\text{He} + 2e^+ + 2v + \gamma$$

The result of this process is that four hydrogen atoms amalgamate to form one atom of helium, two positrons, and two neutrinos. The mass remaining after the reaction has taken place is less than that of the original four hydrogen atoms, with the balance being made up of a photon of extremely high energy gamma radiation. The quantity of energy produced by this annihilation of hydrogen in the solar core is very large, since the conversion factor of hydrogen to energy is the square of the speed of light. It is this relationship between mass and energy that is responsible for the immense energy output of the Sun and other stars.

Emerging from the Sun's core the gamma rays pass through the shell of hot compressed hydrogen where they are transformed by a complex series of collisions and interactions. A little more than two seconds elapses before radiation originating at the centre of the core of the Sun reaches the photosphere. Then, after passing through the Sun's turbulent atmosphere, it emerges into space possessed of the broad spectrum of wavelengths shown in Figure 2.1. All the energy of the solar system, except that resulting from the force of gravitation, is provided by the radiation flux that flows from the Sun at a rate of 3.86×10^{26} W. Such an enormous production of energy consumes the hydrogen fuel of the Sun at the seemingly alarming rate of 4×10^6 tonnes per second. But so large is the Sun's mass that it will continue, even at this rate of loss, to shine with its accustomed brightness for some thousands of millions of years to come.

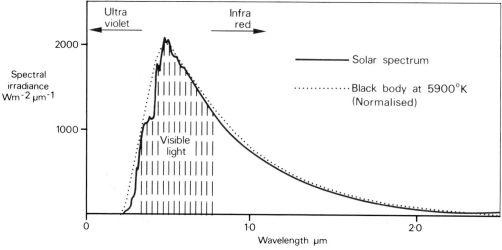

Figure 2.1 The Solar Spectrum

The dotted line on Figure 2.1 describes the electromagnetic spectrum of a perfectly efficient radiant body at a temperature of 5900 K. The closeness of the dotted ideal curve and the solid line of the actual spectrum shows that the photosphere approximates to a black body radiating at about 6000 °C. The differences between them are the result of absorption in the atmosphere of the Sun. Also shown on Figure 2.1 is the visible portion of the Sun's spectrum lying between 0·35 and 0·75 µm. It is worth noting that our eyes are adapted to make use of the most intense portion of the Sun's radiation.

The power density at the Sun's photosphere is 63·4 MW m^{-2}. This number can be put into perspective by saying that the rating of the largest electricity generating station in the United Kingdom, at Longannet in Fife, is 2400 MW. Just 38 m^2 of the surface of the Sun produces energy at the same rate as does Longannet when working at full power.

After a journey of 8·3 minutes the radiation from the Sun arrives at the top of the Earth's atmosphere with a virtually unchanged spectral curve but of course at a much reduced density. The terrestrial solar energy flux, a quantity known as the solar constant, averages 1370 W m^{-2} (Duncan et al., 1982) when measured perpendicularly to the Sun's rays. The solar constant is only 0·002 per cent of the energy density at the photosphere and, due to the elliptical orbit of the Earth about the Sun, it is subject to a variation of about ±3·0 per cent.

THE TRANSFORMATION OF SOLAR ENERGY

During its journey through the Earth's atmosphere the incoming radiation must run a gauntlet of scattering and absorbing substances. These are, in increasing order of size, ions, atoms, molecules, aerosols, dusts, and water droplets. A complicated interaction takes place between the radiation and the constituents of the atmosphere as a result of which the solar spectrum is modified to the curves shown on Figure 2.2. Some of the radiation is absorbed and converted into heat, some is reflected back into space and some is scattered forward to reach the surface of the Earth deflected from its direction of arrival. The middle curve shows the spectrum of arriving radiation in clear weather while the lower represents that reaching the surface through an overcast sky.

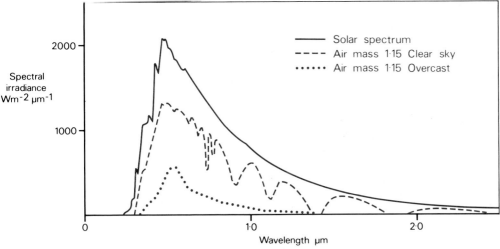

Figure 2.2 Atmospheric Attenuation of the Solar Spectrum

The scale of these effects is proportional to the length of the path through the Earth's atmosphere traversed by the Sun's rays. It is usual to describe this path in terms of air mass. Unit air mass is the distance ZO in Figure 2.3 with the Sun directly overhead while air mass 1·5 corresponds to a solar elevation of 42 degrees. When only 30 degrees above the horizon the rays of the Sun traverse two air masses. In mid-winter in the latitude of London the rays even at noon must penetrate an air mass of 4·5.

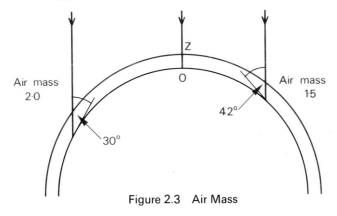

Figure 2.3 Air Mass

The curves shown in Figure 2.2 represent the effect on the incoming radiation of traversing an air mass of 1·15. This corresponds to a solar elevation of 60 degrees, an altitude reached at noon by the Sun in London only between late May and mid July.

Figures 2.4 and 2.5 distinguish the effects of reflection, absorption, and scattering upon the radiation arriving at the top of the Earth's atmosphere when weather conditions are clear, cloudy, or overcast. They are calculated for unit air mass. It will be noted that at ground level the total radiation, referred to as the global radiation, always contains a diffuse component. The direct component, which predominates when the sky is clear, is extinguished in overcast conditions when all solar radiation arriving at the surface is diffuse.

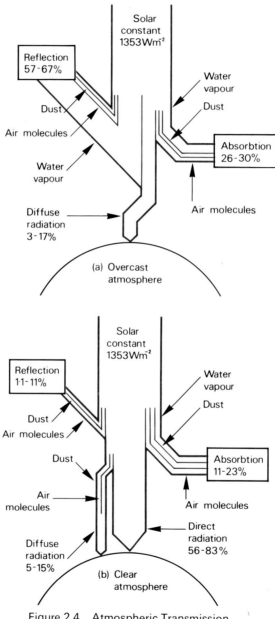

Figure 2.4 Atmospheric Transmission

The daily rotation of the Earth upon its axis is responsible for a regular diurnal variation in the intensity of the radiation received at any location. The detailed pattern of this variation is determined by weather conditions. Figures 2.5(a) to (d) show the solar flux that is typical of days when the sky is clear, partly cloudy, cloudy, and overcast respectively. The diffuse component is that part lying below the lower line. Figure 2.5(a) confirms Figure 2.4(b) in showing that only a small proportion of the insolation arrives at the surface diffused in clear weather. The proportion is larger, about 33 per cent, on a partly cloudy day and rises to 54 per cent when the sky is generally cloudy. Figures 2.4(a) and 2.5(d) both show that all direct radiation is excluded from the surface by a wholly overcast sky.

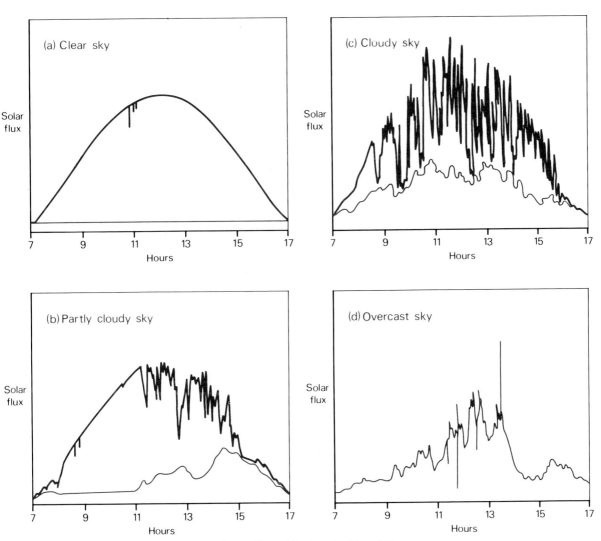

Figure 2.5 Diurnal Variation of Insolation

The intensity of solar radiation arriving at the surface varies with the season of the year as well as the time of day. When Maps 2.8 and 2.14 are measured and compared it is found that insolation in Britain in December is only 9 per cent of that arriving in June. In combination, the various factors affecting insolation levels throughout the year result in the curves shown in Figure 2.6.

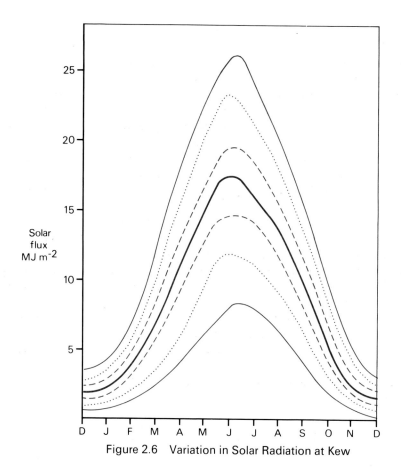

Figure 2.6 Variation in Solar Radiation at Kew

On either side of the heavy line representing the mean daily insolation on a horizontal surface lie dashed lines within which 30-day running averages fall. The four-day running average figures extend between the wider limits of the two dotted lines while the upper and lower firm lines represent the daily amounts of insolation exceeded on 10 per cent and 90 per cent of days respectively. It will be seen that, at the recording station at Kew, months with substantially more or less than the average levels of solar radiation are common and that the proportional variation differs little between summer and winter.

THE MEASUREMENT OF SOLAR RADIATION

Data on solar energy received in the United Kingdom is gathered and published by the National Radiation Centre which forms part of the Meteorological Office. Map 2.1 shows the location of the 31 stations at which systematic measurements of solar radiation levels have been made for at least the past 35 years. From the map it will be seen that the recording stations are too sparse, except perhaps in the Home Counties, to allow synoptic maps to be drawn from the information they supply. Mapping of solar radiation levels must therefore make use of derived data.

There exists a well established correlation between global solar radiation levels and the duration of bright sunshine as a ratio of possible bright sunshine. Because sunshine is a commodity of interest to the leisure and agricultural industries it has been recorded more widely and for a longer period of time than have solar radiation statistics. Map 2.2 shows the distribution of the

132 recording stations for which bright sunshine information is available. Solar radiation levels have been calculated from statistics furnished by these stations and the results, combined with measured data from the stations shown on Map 2.1, form the basis of the isopleths on Maps 2.3 to 2.16. These maps show levels of global solar radiation only.

For purposes of solar energy applications it is often desirable to divide global radiation data into their direct and diffuse components. Unfortunately, only about one-third of the United Kingdom recording stations measure diffuse radiation on a regular basis. The diffuse component can be derived from sunshine duration records but the correlation with those measured values that are available is less good than is the case with global radiation calculations. The dependability of the derived diffuse results is therefore not so well known. No other method of obtaining a synoptic picture of diffuse radiation levels is available, however, and Map 2.17 has been drawn from the calculated data.

The map gives the daily total of diffuse radiation in the United Kingdom averaged over the year. The isopleths follow the same broad pattern as do those of Map 2.15 of global radiation, but the differences are significant and a detailed comparison of the two maps yields the picture shown on Map 2.18. These isopleths describe the proportion of the global radiation that is diffuse. It will be observed that the diffuse component accounts for more than half the solar radiation received everywhere in Britain.

On account of the fact that the intensity of the radiation falling upon a surface varies with the inclination of the surface to the Sun it is necessary to adopt a standard orientation in the presentation of solar statistics. This is done by reducing all solar radiation data to that received by a surface placed horizontally. Data for a horizontal plane can be converted to an equivalent value for any other orientation by calculation. The isopleths drawn on Maps 2.3 to 2.17 give insolation levels on a horizontal surface.

INSOLATION AND SOLAR HEAT

Electromagnetic radiation falling onto a surface can be reflected, absorbed, or transmitted. When radiation from the Sun encounters an object a certain proportion, depending upon the nature of the surface, will be absorbed and appear in the object in the form of heat. A matt black surface will absorb nearly all of the radiation falling upon it and an object possessing such a surface will therefore gain heat rapidly. A polished metal surface, on the other hand, will reflect more than 90 per cent of the incident radiation. As an object gains heat its temperature will rise and as a result it will itself become an emitter of radiation. At 300 K, or 27 °C, the surface of the object will emit the thermal radiation spectrum shown on Figure 2.7. Most objects on the Earth's surface are maintained by solar radiation at a temperature between the freezing and boiling points of water, the temperature range within which the processes of life can function.

It will be observed that the wavelength of solar radiation does not exceed 2·0 μm, while virtually the whole of the thermal spectrum lies beyond 3 μm. The separation of the two spectra is exploited in the design of solar collection devices.

It is fortunate that so common a material as glass is transparent to solar radiation while remaining opaque to longer wavelength thermal radiation. Superimposed upon the two spectra shown in Figure 2.8 is the transmittance of water-white glass. It can be seen that a pane of this glass will readily transmit solar radiation but will block the passage of radiation at thermal wavelengths. This is the basis of what is known as the greenhouse effect, by the operation of which solar radiation can be trapped in the form of heat behind a sheet of glass. The greenhouse effect is responsible for the fact that buildings can gain and retain much heat from the sun shining through windows, glass walls, and roofs or other glazed openings. This property of glass is also utilized in the design of many solar collector devices.

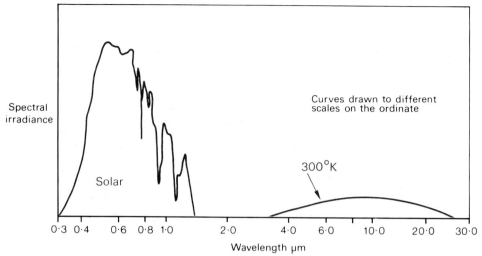

Figure 2.7 Solar and Thermal Radiation Spectra

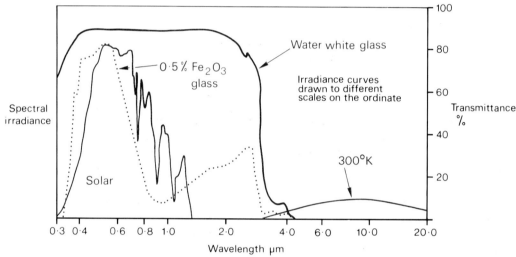

Figure 2.8 Transmittance of Glasses

The transmission curve of glass containing 0·5 per cent ferrous oxide is also shown on Figure 2.8. Glass of this type is cheaper and more common than iron free glass but because it is opaque to a significant proportion of the Sun's radiation it does not discriminate so well between the two spectra.

Solar radiation collection devices are usually finished with a matt black surface to facilitate absorption of the sun's radiation. Their temperatures will in consequence often rise to above the boiling point of water in sunny weather. The emission spectrum will then be near the 400 K, or 127 °C, curve shown on Figure 2.9. It is evident from this diagram that the incoming energy can be most effectively absorbed and retained by a surface which combines a high absorptivity in the region of the solar spectrum with a low emissivity at thermal wavelengths. The curve of an ideal selective surface, expressed in terms of its reflectance, is shown on the graph together with the performance that can be achieved by a metal surface treated with a commercial finish of the black chrome type.

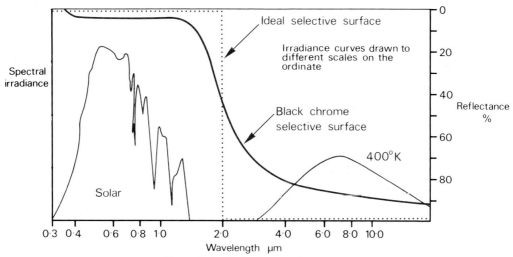

Figure 2.9 Selective Surfaces

The performance of a solar collector is greatly influenced by the efficiency with which its surface enhances the absorption of solar radiation while inhibiting the emission of energy in the thermal part of the electromagnetic spectrum.

THE MAPS

Maps 2.3 to 2.15 show daily totals of global solar radiation received in each month of the year together with an annual average daily value. In the United Kingdom the latitudinal range of 11·5 degrees results in a marked decrease in radiation levels from south to north. The relative decrease across the country is greater in winter, when the December insolation levels vary by a factor of six, than in summer when the variation in June levels is only 1·3. The factor averaged over the year between the Channel Islands and Shetland is 1·5 and is primarily the effect of the greater air mass traversed by the rays of the Sun in reaching more northern latitudes. When the monthly data is summed over the year the isopleths drawn on Map 2.16 emerge, where the total quantity of solar radiation received each year is described.

At all times of the year the solar radiation falling on the high ground of the Welsh Mountains, the Pennines, the Southern Uplands, and the Highlands of Scotland is less than in adjacent lowland areas. The cloudy skies characteristic of mountain country are responsible for this effect. It may be noted that neither the Scottish Lowland valley nor the English Lake District are large enough to create a solar radiation regime of their own. The Home Counties, where one-quarter of the population of England and Wales live, are in all months an area of relatively low insolation on account of the atmospheric pollution arising from the metropolis of London.

The diffuse solar radiation regime described on Map 2.17 follows the general pattern of the global radiation statistics, with a diminution from south to north and with low levels of radiation on high ground and in southeastern England. A high proportion of the solar radiation in the industrial midlands of England is diffuse and amounts to 66 per cent of the global figure on the borders of Staffordshire and Derbyshire. Map 2.18, which gives the diffuse component as a percentage of the global level, shows that in no part of the country does direct sunlight account for as much as half the energy received. In Jersey 52 per cent is diffuse while at the northern extremity of Shetland more than two-thirds of solar radiation arrives in diffuse form. It is for this reason that solar energy collection devices for use in the British climate must be capable of exploiting diffuse as well as direct sunlight.

THE SIZE OF THE RESOURCE

It has been shown that the solar radiation energy flux arriving in the United Kingdom is relatively low. Even a solar energy flux as high as 1 kW m^{-2}, a figure rarely exceeded in the British climate, is a thousand times less than the performance achieved by modern steam boilers in which heat fluxes as large as 1 MW m^{-2} are commonplace. Since only about 45 per cent of this low intensity radiation reaches the surface as a direct beam it follows that practical solar energy collectors must, in Britain, be adapted to accept radiation which is of low intensity and of a diffuse nature. The device most commonly employed for this purpose is the flat plate solar collector.

A flat plate solar collector may be likened to a conventional central heating radiator but designed to work in reverse, that is to work as a heat sink rather than as a source of heat. In service the plate is positioned to expose the maximum area to the incoming solar radiation which, when intercepted by striking the plate's surface, is converted into heat. Water or air channels in the collector serve to gather the heat and to transport it elsewhere for use. All but very low temperature collectors are furnished with a glass cover plate to exploit the greenhouse effect and their collecting surface is treated with a selectively absorbing coating.

A flat plate device 'sees' an entire hemisphere and, limited only by the reflectivity of the glass cover, can absorb radiation arriving from any direction within its field of view. Collectors of this type are therefore well adapted to receive diffuse radiation. Because of their large surface area it is characteristic of flat plate collectors that their efficiency declines sharply when the operating temperature rises above about $70\,°C$. They are therefore best regarded as providers of low grade thermal energy.

Heat at $70\,°C$ or less is useful for the space heating of buildings and for certain agricultural and leisure purposes. Fish farming, horticulture, and swimming pool heating are minor energy items in the United Kingdom but heating the space within buildings and supplying them with hot water accounts for 32 per cent of the country's entire requirement for energy. Figure 2.10 shows the energy used by the principal sectors of the economy in 1979 and also lists by sector the amount used for space and water heating.

	Whole sector PJ	Heating and HW only PJ
Industrial	2445	465
Domestic	1741	1274
Transport	1481	
Other	843	371
Total	6510	2110

Figure 2.10   Energy Consumption in the UK Economy 1979

Solar collector panels are not always indispensible for supplying a building with heat from the sun. Every building, as it intercepts and absorbs solar radiation, gains heat and so, however unintentionally, acts as a solar collector. By taking appropriate measures a building can be designed to maximize these effects. The description 'passive solar design' is often used to distinguish the architectural approach to the utilization of solar energy from 'active' systems employing collector panels or other mechanical devices.

Only a few of the small number of solar buildings so far erected in Britain have been studied with sufficient care to produce reliable data on their solar and thermal performance. Some information has, however, been gathered from two public housing projects completed in recent years. The solar house at Milton Keynes, built in 1974, is heated by an active solar system employing 40 m² of flat plate collector panels. It has been found that the contribution made by the sun to the building's annual energy requirement for space and water heating is 63 per cent (Hodges and Horton, 1979).

Further north, at Bebington near Birkenhead, a terrace of houses was built in 1977 to a design which maximizes their ability to absorb solar radiation by passive means without the aid of solar collector panels. These solar houses use only 50 per cent of the energy consumed by an adjoining terrace of similar conventionally-heated houses built at the same time as an experimental control (Justin et al., 1980). Since it is likely that fully developed solar buildings, particularly those of a larger size than a house, will make use of both active and passive solar techniques, it may be concluded that the direct use of solar radiation could in due course provide 57 per cent of the energy needed to heat buildings in Britain. This would amount to an annual total of 1203 PJ, or 18 per cent of the entire energy consumption of the United Kingdom economy in 1979.

THE SOLAR ENERGY REGIME OF NORTH AMERICA

From the southern tip of Texas to Cape Columbia at the northern extremity of Ellesmere Island in the Canadian arctic archipelago is a distance of some 6170 km spanning 55·5 degrees of latitude. Since the intensity of solar radiation at the surface of the Earth varies with the angle of incidence of the Sun's rays it is to be expected that insolation levels will vary greatly across such a large latitudinal range. In fact the annual average quantity of insolation received at Cape Columbia is only one-third that arriving at El Paso on the border of Mexico and New Mexico. The pattern of this variation is not, however, uniform.

The elevated country occupying the middle section of the border with Mexico is the sunniest region of the United States. It lies at the centre of that portion of the southwestern states sometimes known as the 'sunbowl'. The high country of the main chain of the Rocky Mountains extending northward through Colorado, Wyoming, and Montana experiences the cloudy skies characteristic of mountain regions. Insolation levels in the Rockies are comparatively high, however, because the air in high country is rarified and clean. The larger proportion of the incoming radiation that is transmitted to the ground through the clean air of the mountains more than compensates for that lost to space by reflection from clouds. As a result the $6·5 \text{ GJ m}^{-2}$ isopleth reaches as far north as central Wyoming while the $6·0 \text{ GJ m}^{-2}$ contour extends almost to the Canadian border.

The isopleths drawn on Map 2.19 show a relative lessening of the intensity of solar radiation across the plateau of Idaho and the northern part of Utah. To the west the mountains of the Sierra Nevada and the Cascades move the isopleths north again until, nearing the west coast, they are deflected to the south by cloudier skies near the Pacific Ocean. Further north on the west coast of Canada and in southern Alaska the isopleths experience an even more pronounced southward shift because the prevailing westerly and southwesterly winds make this an area of unusual cloudiness and very heavy rainfall.

The Prairies of the United States, drained by the Mississippi River, are cloudier than the Great Plains which border the eastern side of the Rocky Mountains. As a result the $6·5 \text{ GJ m}^{-2}$ isopleth is shifted from central Wyoming south and east to the coast of Louisiana on the Gulf of Mexico. The influence of the high cloud levels of the Appalachian Mountains is perceptable in the trend of the isopleths over the eastern United States.

The effects of southerly or southwesterly prevailing winds and of a warm ocean current flowing north from the Caribbean Sea combine to produce relatively high levels of insolation near the Atlantic seaboard as far north as the coast of Maine. In upstate New York the cloudy skies characteristic of the Adirondack Mountains produce a large area where annual insolation amounts to only 4·5 GJ m^{-2}.

The southern part of the Canadian provinces of Alberta and Saskatchewan benefit from an extension northward as far as Edmonton and Lake Winnipeg of the 4·75 GJ m^{-2} isopleth. In fact, the agriculturally productive part of the Canadian Great Plains is almost defined by this contour. Further eastward the cloudier weather associated with Hudson Bay displaces the isopleths southward before they tend east and then southeast to the Atlantic Ocean. Insolation levels are higher in the Canadian arctic archipelago, where the 3·0 GJ m^{-2} isopleth crosses Ellesmere Island in 80 degrees north latitude, than they are in Greenland or over the Beaufort Sea to the north of Alaska. Solar radiation levels in Alaska are remarkably uniform with almost the entire state lying between the values of 3·0 and 3·5 GJ m^{-2}.

Map 2.16 of Britain and Map 2.19 of North America have been measured to obtain the data upon which Figures 2.11 and 2.12 are based. The graphs plot the proportions of the land areas of the two regions against the levels of insolation that they receive. The areas beneath the curves are therefore proportional to the sunniness of the respective countries. The slope of the graph gives an indication of the variation occurring in the levels of solar radiation. In the United Kingdom insolation varies by a factor of 1·5 following from the fact that a group of islands 1300 km long lying across the path of the North Atlantic ocean current will experience a relatively uniform solar radiation regime. The North American continent, however, extending from the high arctic nearly to the tropics, has a correspondingly large range of insolation levels. The slope of the line in Figure 2.11 quantifies this phenomenon.

Figure 2.12 distinguishes the annual insolation levels of the conterminous United States, Canada, and Alaska. The United States is shown to be 50 per cent sunnier, and Alaska 20 per cent less sunny, than Canada. The curve of the United Kingdom solar radiation regime lies between those of Canada and Alaska.

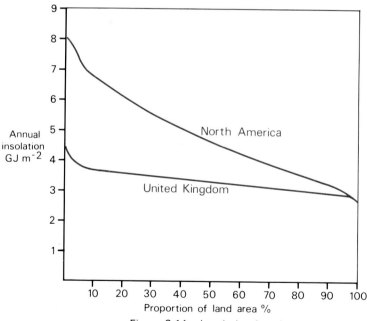

Figure 2.11 Insolation Levels

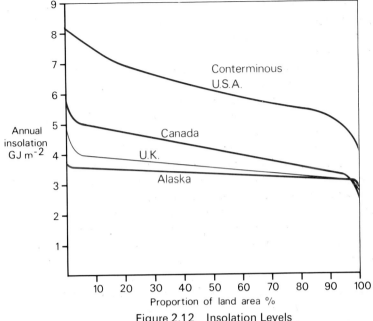

Figure 2.12   Insolation Levels

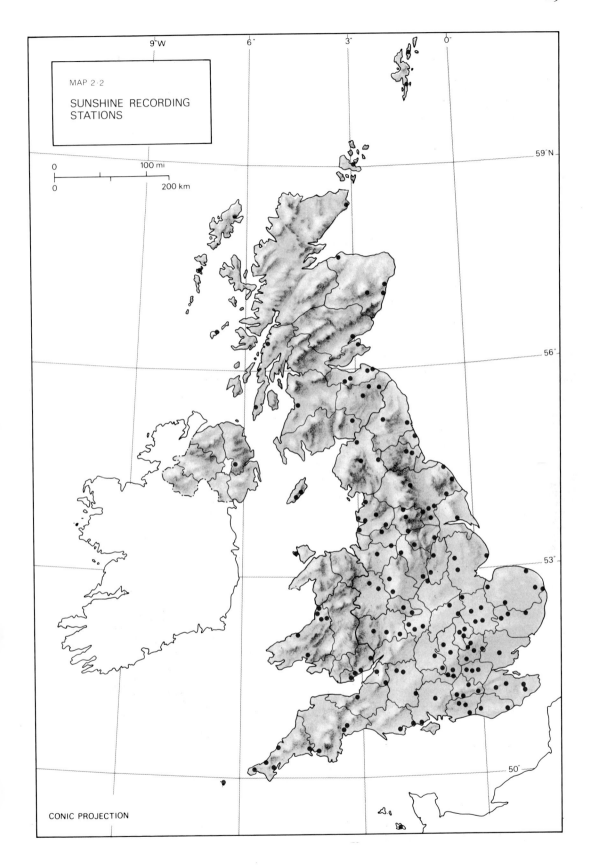

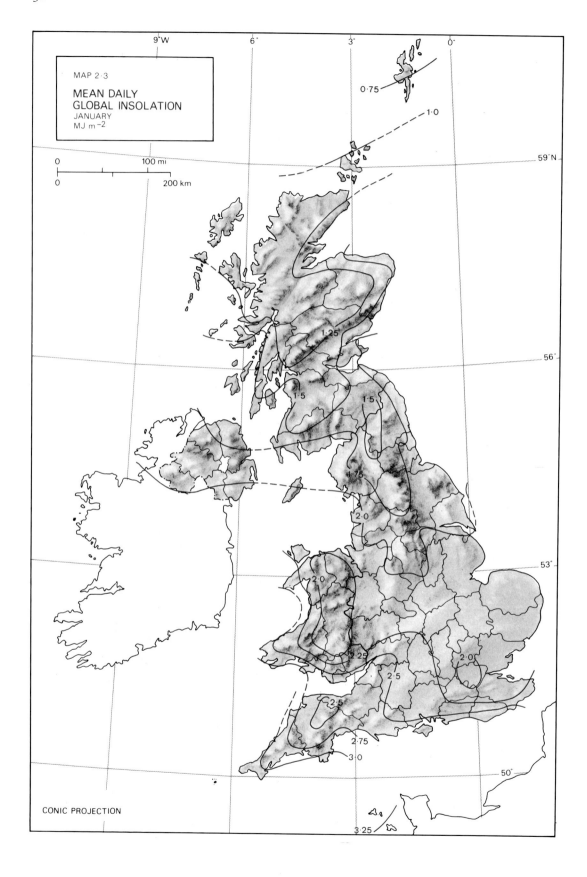

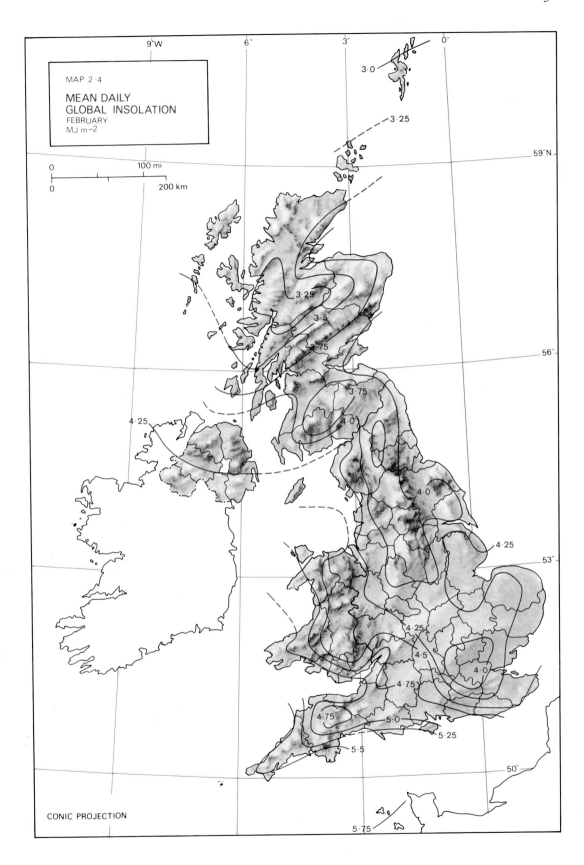

MAP 2·4
MEAN DAILY GLOBAL INSOLATION
FEBRUARY
MJ m−2

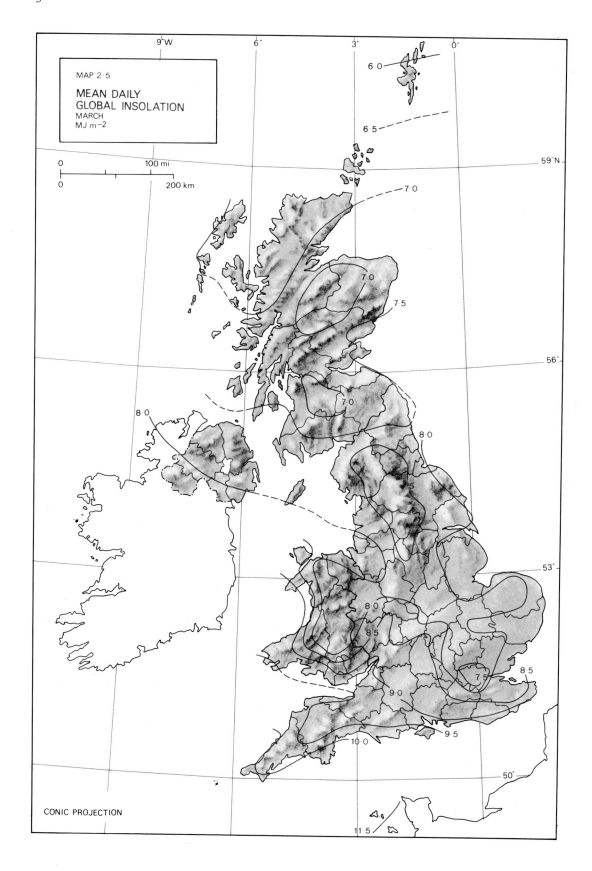

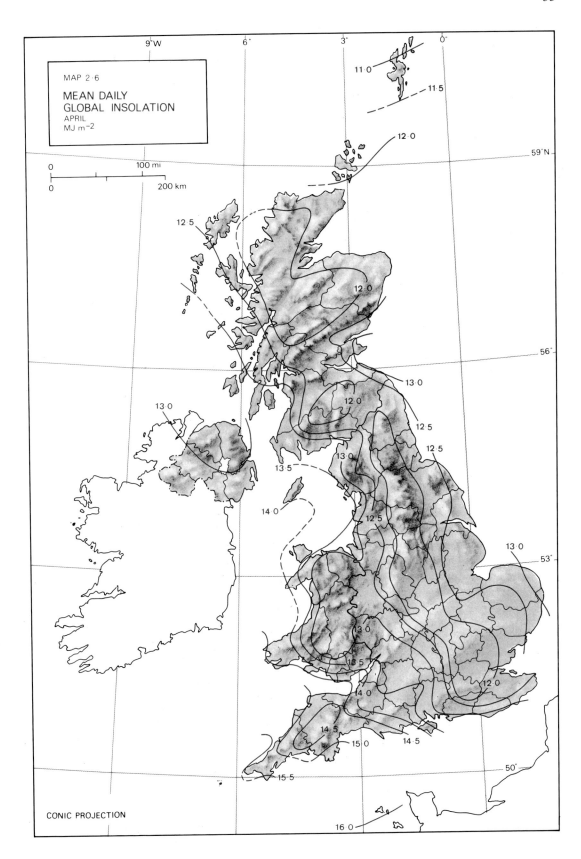

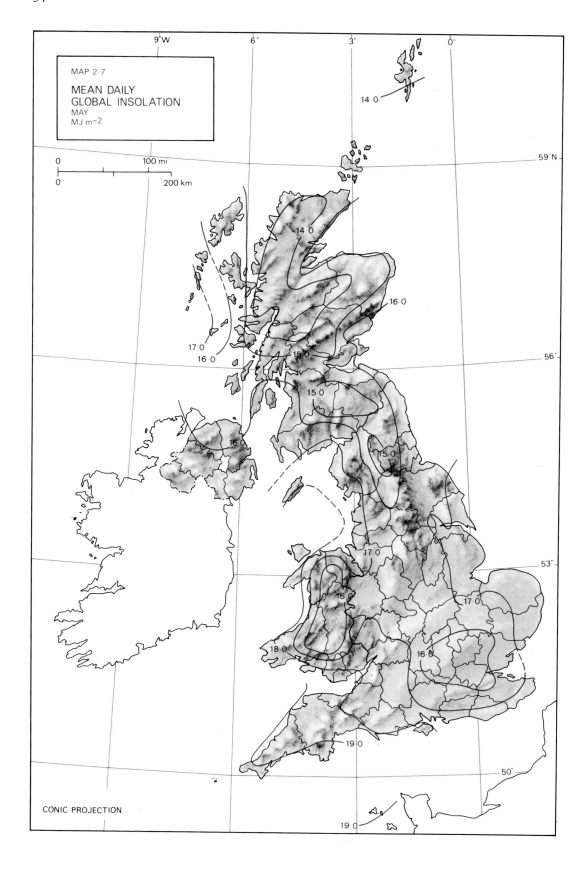

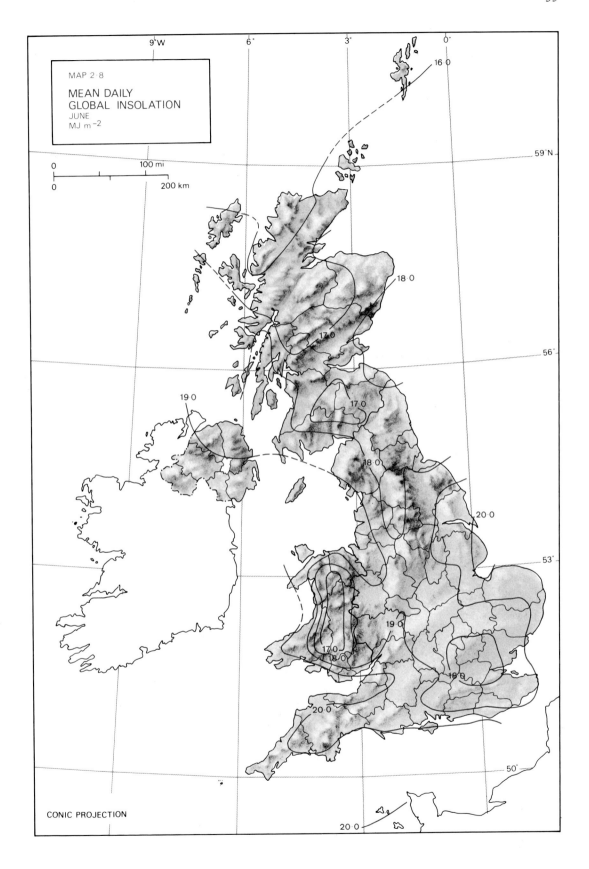

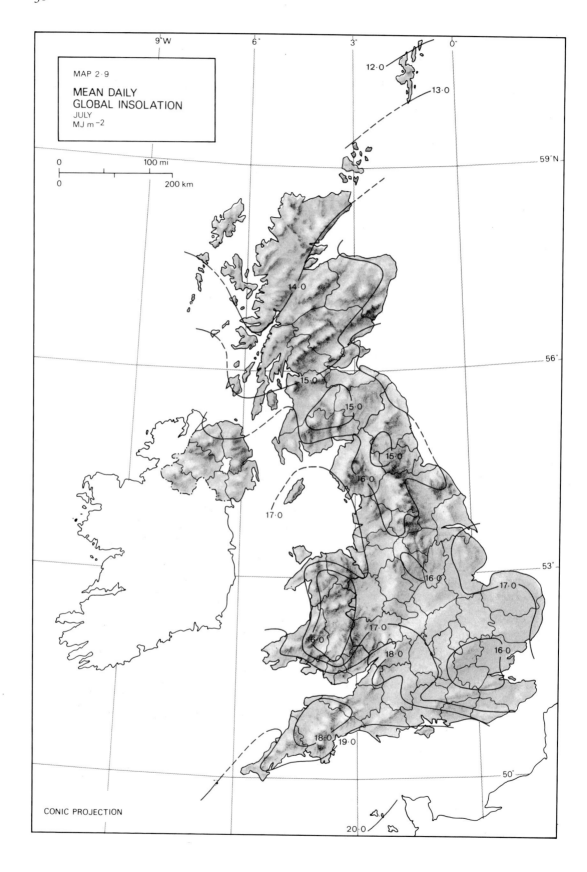

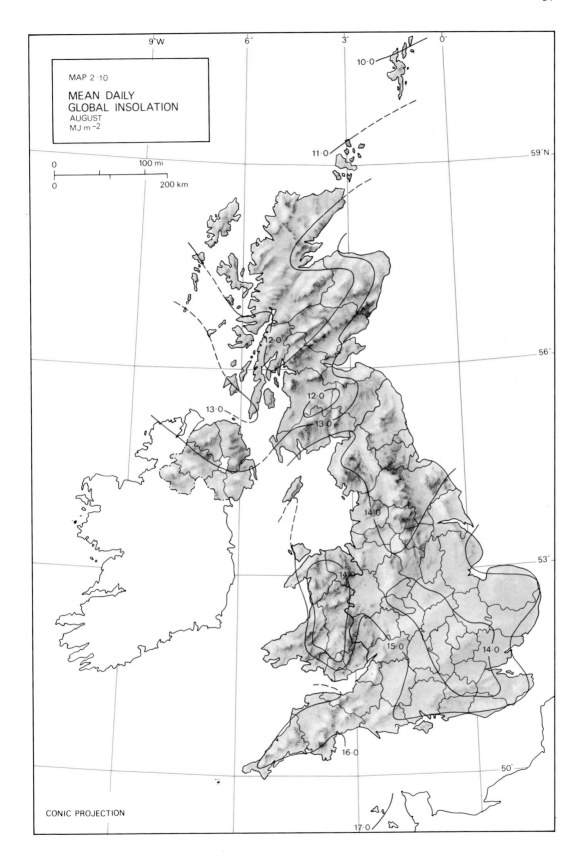

MAP 2·10
MEAN DAILY GLOBAL INSOLATION
AUGUST
MJ m⁻²

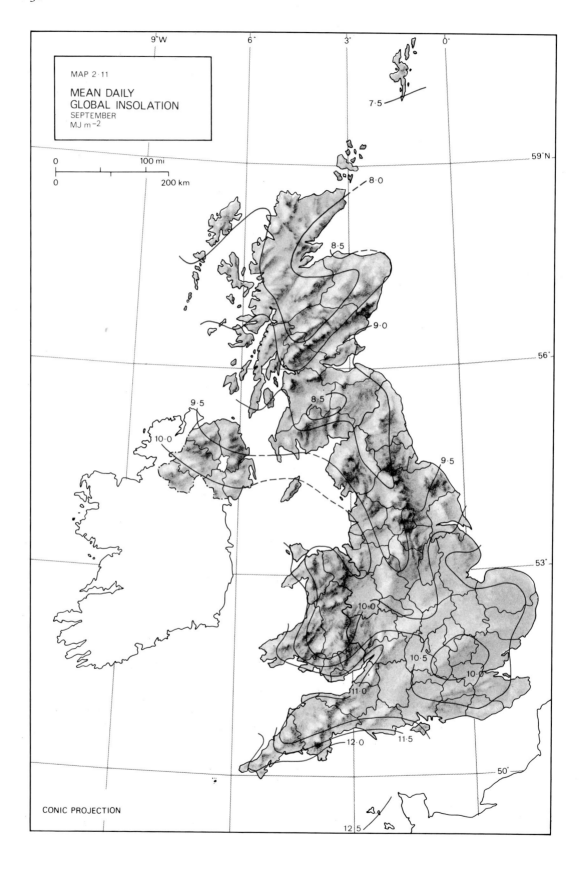

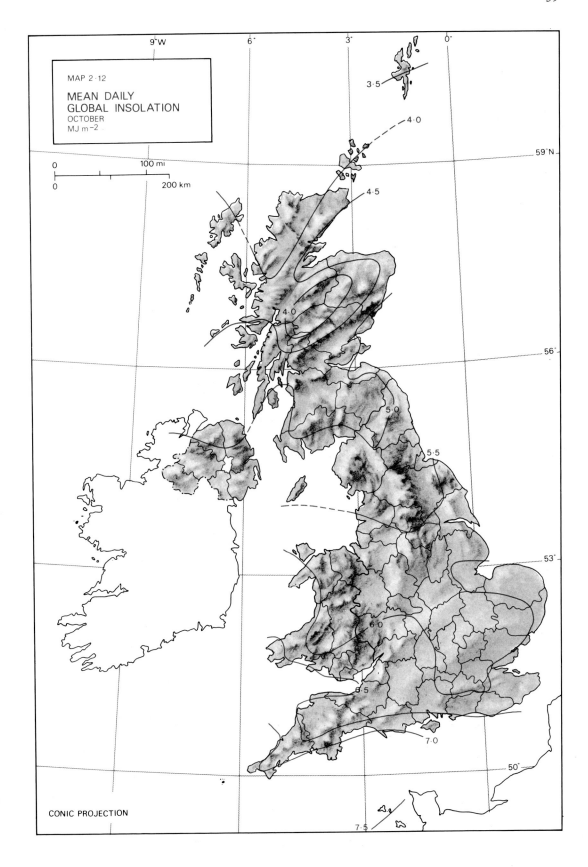

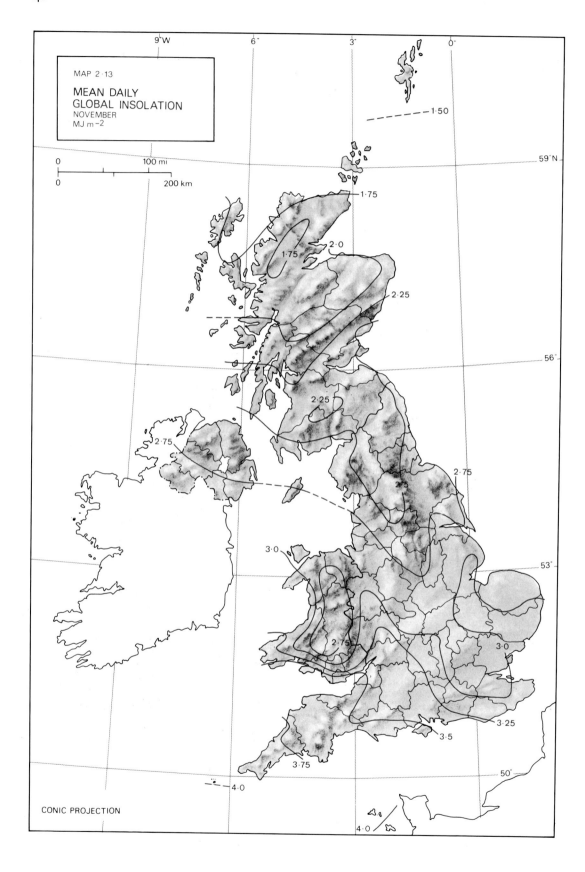

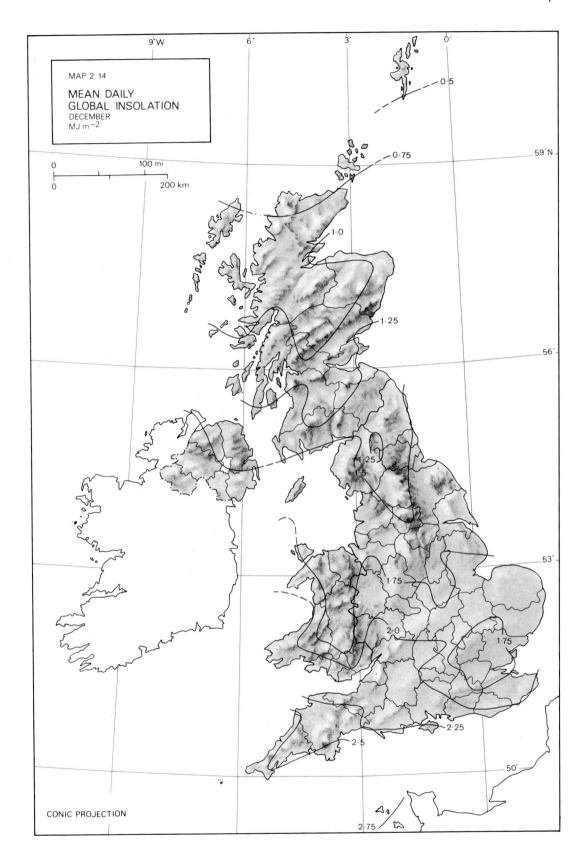

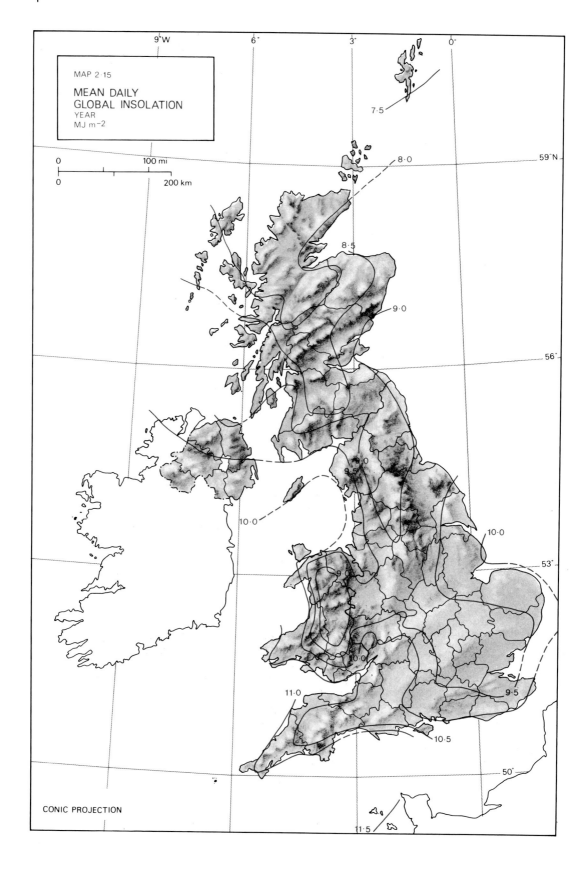

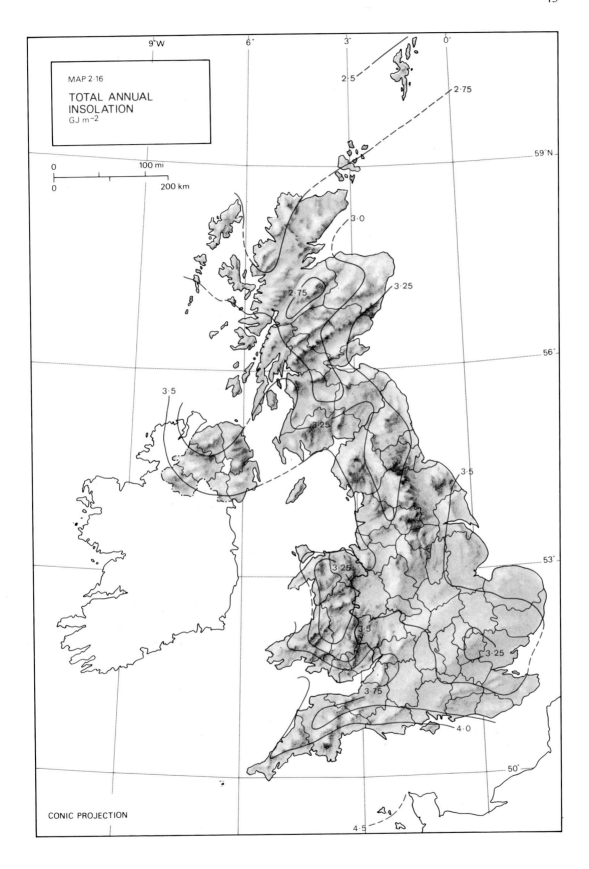

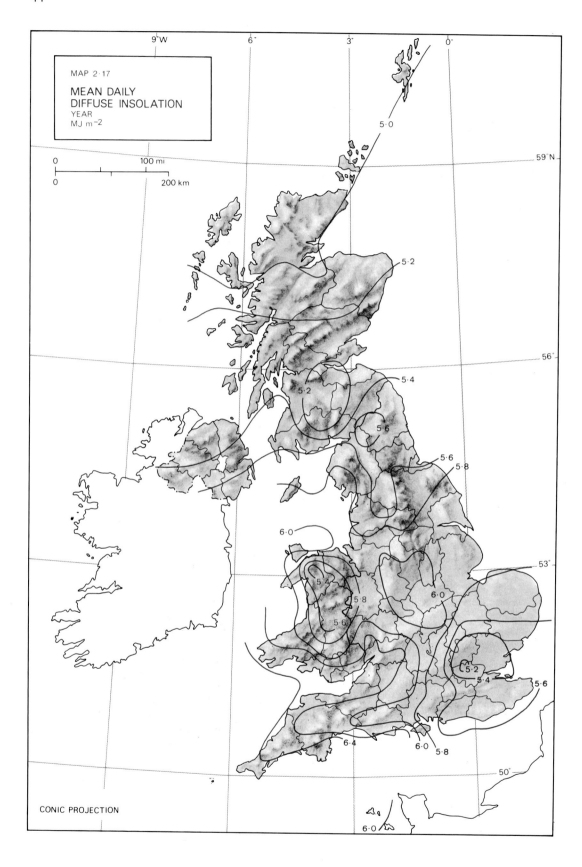

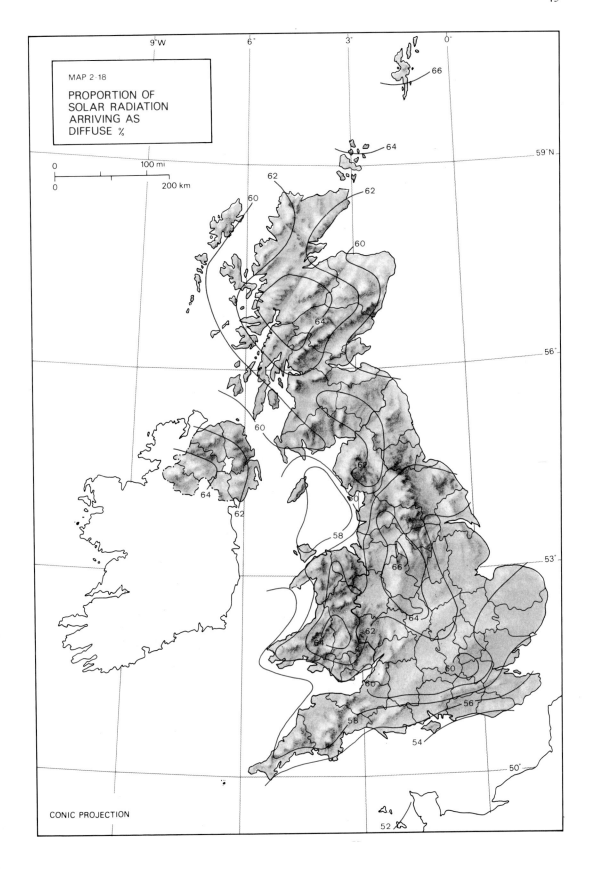

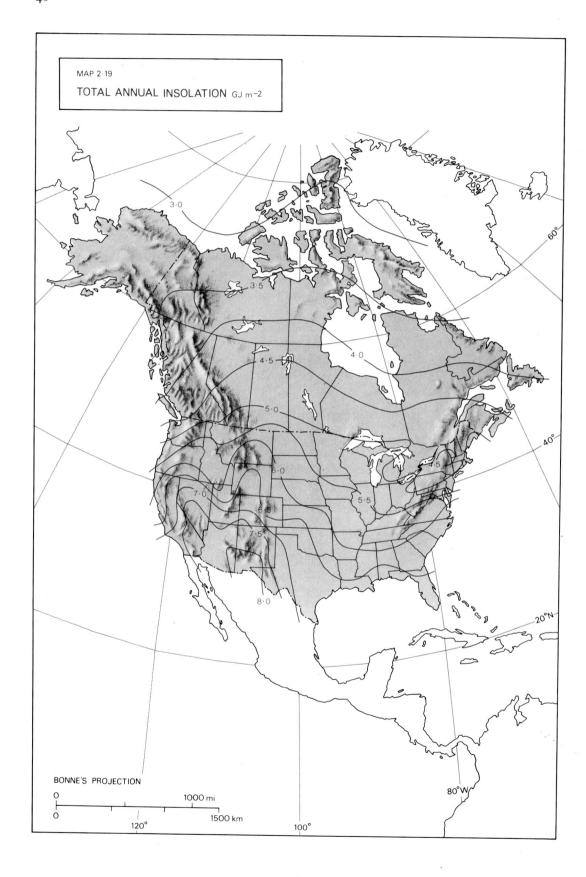

Chapter Three
Wind Energy

THE NATURE OF THE WIND

So great is the Earth's income of solar radiation that a small proportion of it, less than 2 per cent of the total, is sufficient to power all the wind, waves, and ocean currents of the world. One effect among many of the operation of the huge heat engine which we perceive as the weather is the creation of a constantly varying pattern of high and low atmospheric pressures over the surface of the globe. The mechanism by which these worldwide variations in pressure arise and interact is extremely complicated and only partly understood. It is sufficient for the present purpose to know that the Earth's atmosphere is constantly in migration from places where pressure is high to areas of low pressure in a never-ending endeavour to arrive at equilibrium.

Wind is thus the atmosphere in motion. The air constituting the atmosphere has a small but measurable mass and in motion it possesses the energy of all moving things, known as kinetic energy. This energy has been tapped by mankind for centuries and for most of human history it has, with the water of flowing rivers, been his only non-animal source of power. Today, when fossil fuels are beginning to become scarce, wind power is once again receiving attention as an important renewable energy resource.

THE QUANTIFICATION OF WIND ENERGY

The kinetic energy imparted by a moving mass of air in unit time may be found from the following formula (Golding, 1976):

$$E = 0.5 \rho A V^3 \, \text{J s}^{-1}$$

where ρ = air density (kg m^{-3})
V = air velocity (m s^{-1})
A = area through which wind passes normally (m^2)

A useful simplification of this formula can be made by substituting the value of $1.201 \, \text{kg m}^{-3}$ for the density of air in a standard atmosphere where the temperature is $17\,°\text{C}$ and pressure equals 1000 millibars. Then

$$E = 0.6 A V^3 \, \text{W}$$

It will be noted that the kinetic energy of moving air is proportional to the cube of its velocity.

Although the energy of the wind is reduced by the decrease of atmospheric pressure with height, and will be affected by variations in the water content of the air, these deviations from a standard atmosphere are too small to have any practical effect on the magnitude of wind energy availability in the United Kingdom. The simplified formula may therefore be used in wind energy assessments in this country.

WIND ENERGY STATISTICS

Information on wind speed, variability, and direction in the United Kingdom has for many years been systematically collected and recorded by the Meteorological Office. This data is processed by computer and information on other aspects of the wind regime, such as the frequency of calms and gales and maximum expected gust strengths, is also available on a nationwide scale from the database.

It is unfortunate, from the point of view of energy studies, that the efforts of the Meteorological Office have so far been mainly directed towards predicting wind loads on buildings, bridges, and other structures as well as providing information to the forestry and agricultural industries on wind damage risks. The siting of wind recording stations reflects this area of interest and they are not well placed for collecting data on wind as an energy resource. The location of 37 of the some 112 recording stations in the United Kingdom is shown on Map 3.4 from which it will be seen that nearly all are sited in inhabited areas below an altitude of 250 m. In order to arrive at comparable results the recording anemometers are usually sited well away from obstacles to the flow of the wind and are positioned at a height above ground of 10 m. Average wind speeds shown on Map 3.1 therefore refer to unobstructed lowland rural sites.

Future research can be expected to reveal that wind speeds in upland areas are higher than is shown on Map 3.1. Hill sites in the Pennines, the Cambrian Mountains, the Southern Uplands of Scotland, and the Scottish Highlands will probably be found to have average wind speeds of as much as 50 per cent greater than is shown. Furthermore, it is likely that sites at sea as little as 5 km from the coast will experience considerably stronger winds than nearby locations ashore where friction and obstructions retard the motion of the air. It can be expected that the isotachs, or lines of equal wind speed, on Map 3.1 will need substantial revision when data from offshore and mountain sites becomes available. In the meantime the map should not be used as a source of accurate wind speeds in hilly and coastal regions.

VARIATION OF WIND SPEED WITH HEIGHT

Below an altitude of 500 m the speed of the wind is reduced by friction against the surface of the Earth, with the slowing proportional to the roughness of the ground over which the air passes. Smooth water creates the least friction, while the greatest retardation is caused by buildings in built-up areas. Standard data, applicable to flat unobstructed sites at a height of 10 m, may be converted to allow for other conditions by use of the graph in Figure 3.1. It enables approximate corrections to be made to standard data when the site of interest is obstructed or when it is proposed to instal a wind machine at a height other than 10 m.

The curves in Figure 3.1 are obtained from the following formula:

$$V_h = V_{10}/(10/h)^n \, \text{m s}^{-1}$$

where V_h = wind velocity at height h above ground (m s^{-1})
V_{10} = wind velocity at 10 m above ground (m s^{-1})
h = height above ground (m)
n = experimentally determined exponent:
  0·16 at flat unobstructed rural sites
  0·30 at obstructed rural or normal suburban sites
  0·40 at flat urban sites

The rate of variation of wind speed with height, as well as air speed at 10 m, is affected by the topography of a particular site, the precise nature of the ground surface, and, since trees lose their leaves in winter, the season of the year. These effects cannot, in the present state of knowledge, be predicted accurately.

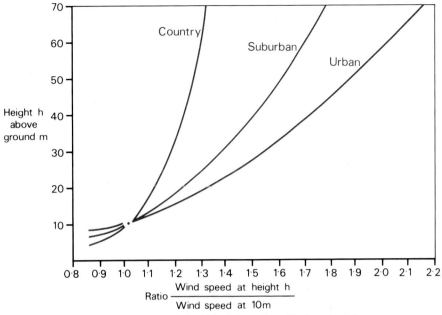

Figure 3.1 Wind Speeds at Various Heights

ANNUAL VARIATION OF WIND ENERGY

The average distribution of the energy of the wind throughout the year in the United Kingdom is given in Figure 3.2. Monthly figures for wind energy at a particular site may be obtained by applying these percentages to the appropriate value extracted from Map 3.2.

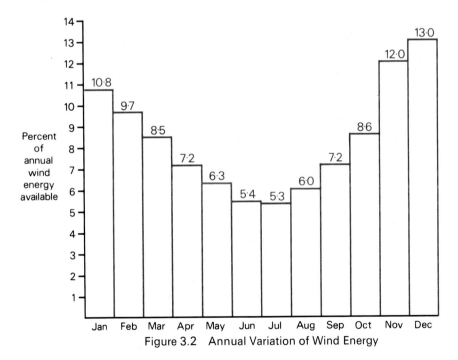

Figure 3.2 Annual Variation of Wind Energy

THE MAPS

It has been seen that the kinetic energy of the wind is proportional to the cube of its velocity. Map 3.1, which shows annual mean wind speeds in the United Kingdom, therefore provides a rough guide to the national distribution of the wind energy resource. Highest average speeds are encountered in the Hebrides and the lowest are in the London area. Low figures for central Wales, the Pennines, and the upland parts of Scotland should, for the reasons already discussed, be treated with caution.

Map 3.2 is obtained by taking into account the velocity characteristics of the wind. Winds blow at different speeds for various lengths of time. By applying the wind energy formula described earlier to each wind speed for the length of time it blows the total annual wind energy of a site can be found by summation. When the results are plotted the pattern of isopleths shown on Map 3.2 emerges. It can be seen that the wind energy available in the Hebrides is eight times that obtainable in the central parts of the midlands and south of England. This disparity reflects the continuous nature of the wind in northwest Scotland where calm weather is rare.

At a height of more than 500 m in flat open country the wind blows at a steady strength and in a constant direction. Nearer the ground, however, friction and obstructions cause the wind to become turbulent. Turbulence is responsible for the proverbial fickleness of the wind and it is a feature of the wind regime that is disguised by average wind speed statistics. In windy weather turbulence produces the large variations in wind speed and direction which are perceived as gusts. Buildings, windmills, and other structures must be capable of surviving the expected gust strength of the wind. The isotachs on Map 3.3 show wind speeds not likely to be exceeded at a height of 10 m for longer than 3 s more than once in 50 years. Design wind speeds may be derived from these data by normal engineering methods.

WIND SPEED DISTRIBUTION GRAPHS

Two separate locations may, while having the same average wind speed, possess very different wind regimes. Differing proportions of calm and of light, moderate, and strong wind, although averaging to the same figure, will result in quite different operating conditions for wind energy machines. The wind conditions of a site are best described in the form of a wind speed distribution graph. The curve of such a graph is obtained by plotting wind speed on the ordinate against on the abscissa the number of hours in the year at which the wind blows at that strength. The shape of the distribution curve is a good guide to the wind energy potential of the site.

Clearly the windiest sites such as Costa Hill, Mynedd Analog, or Rhossili Down are those with the largest area under the curve. The detailed character of a site may, however, be perceived by examining its speed distribution curve in the manner described in Figure 3.3.

Map 3.4 shows the location of 37 wind recording sites in the United Kingdom for which velocity distribution graphs are supplied in Figures 3.6 to 3.10. It will be noted that the windiest sites are those on hilly ground near the sea, a combination of circumstances more commonly encountered on western than eastern coasts. Future wind resource surveys are likely to reveal many promising wind energy sites on high ground in the Highlands and Southern Uplands of Scotland, the Pennines, the Cambrian Mountains, and on the moorland areas of the West Country. The fact that the progress of the wind over the sea is little retarded by friction and turbulance points to shoal waters near coasts, such as the Owers and Royal Sovereign Banks in the English Channel or the many sandbanks in the Thames Estuary, as likely offshore locations for wind energy machines singly or in groups.

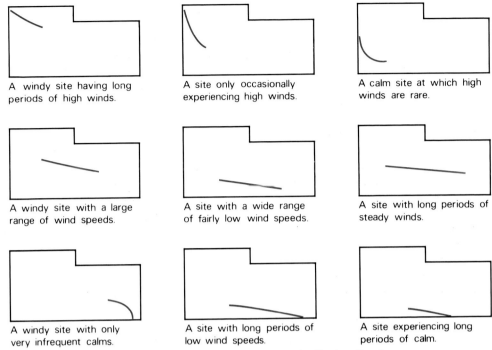

Figure 3.3 Characteristic Velocity Distribution Graphs

THE SIZE OF THE WIND ENERGY RESOURCE

In the summer of 1979 a medium-sized electricity generating windmill, or aerogenerator, was brought into use at Aldborough in the West Riding of Yorkshire. It is a horizontal axis three-bladed machine 17 m in diameter and is a typical example of modern windmill design.

An idea of the size of the wind energy resource may be obtained by measuring the output of the Aldborough aerogenerator against the annual consumption of electricity in the United Kingdom. In 1979 some 848 PJ of electricity was delivered to consumers. If the Aldborough machine were erected upon a moderately favourable wind site such as that at Fleetwood it would produce 474·8 GJ in a year. Simple division produces an hypothetical requirement for 1,786,015 machines if the whole British requirement for electric power were to be obtained from the wind resource.

A more plausible picture emerges from a consideration of the performance of the 91 m diameter aerogenerator recently completed in the United States by the General Electric Company. If this machine were to be erected on a favourable wind site such as Rhossili Down at the western tip of the Gower peninsula in South Wales it would generate 55·5 TJ of electrical energy every year. The United Kingdom economy could be supplied with all its electricity by 15,279 windmills of this size. A round number of 10,000 GEC aerogenerators would furnish 555 PJ a year. Although the engineering, organizational, and environmental problems entailed in deploying 10,000 large machines are formidable, it remains that the United Kingdom wind regime could be tapped to provide 65 per cent of our present requirement for electricity.

THE WIND REGIME OF NORTH AMERICA

As in the United Kingdom the winds of the North American continent are generally found to be stronger in coastal areas. The isotachs on Map 3.5 show average wind speeds of $7\,\mathrm{m\,s^{-1}}$ on the west coast of Queen Charlotte Island and off the east coasts of Nova Scotia, Newfoundland, and Labrador. In the approaches to Frobisher Bay on Baffin Island the average wind speed reaches $8\,\mathrm{m\,s^{-1}}$, a figure that can only be matched in the British Isles on the west coast of the Outer Hebrides.

Winds decrease in strength at lower latitudes on the east coast of the United States falling to $4\,\mathrm{m\,s^{-1}}$ in southern Florida. The coast of Texas is slightly windier but the west coast of the United States as far north as Puget Sound is remarkably calm with averages rising to only $4\,\mathrm{m\,s^{-1}}$ in the San Francisco area and in the Sacramento valley. The western archipelago of Canada and also the coasts of Alaska are windy but in the Canadian arctic archipelago and in Baffin Bay wind speeds are as low as $3\,\mathrm{m\,s^{-1}}$.

Inland, the North American wind regime exhibits higher wind speeds in the heart of the continent. The $5\,\mathrm{m\,s^{-1}}$ isotach extends from the Canadian arctic south to the coast of Texas, with slightly calmer areas in southern Canada and the southern part of Texas. Hudson Bay and the province of Keewatin experience wind speeds nearly as high as those occurring on the Atlantic and Pacific coast of Canada.

In the Canadian part of the Rocky Mountains and in the central parts of Alaska winds are light with averages of less than $2\,\mathrm{m\,s^{-1}}$ on the border of Alaska and Canada. The Rocky Mountains of the United States, although windier than those of Canada, are calmer than the Great Plains to the east. It is likely that further investigations will show that these low average wind speeds are typical only of the inhabited valley country of the Rockies and that high winds are to be found in elevated terrain.

A large area covering the southern part of the Appalachian Mountains and the states of Virginia, Kentucky, Mississippi, Alabama, southern Georgia, and western Florida exhibit average wind speeds of only $3\,\mathrm{m\,s^{-1}}$. This and the west coast are the calmest parts of the United States. Were wind records available from higher ground in the Appalachians they would probably, as in the Rockies, show that good wind energy sites are to be found in elevated areas of otherwise mostly calm districts.

The graph shown in Figure 3.4 has been obtained by measuring Maps 3.1 and 3.5 in order to be able to compare and illustrate the relative windiness of the United Kingdom and North America. Average wind speeds are plotted against the percentage of the land and are of the two

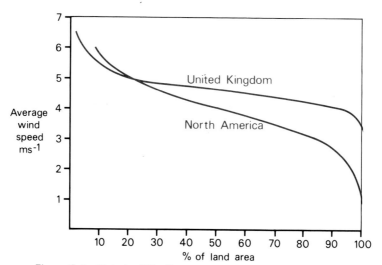

Figure 3.4 Relative Windiness of United Kingdom and North America

maps over which the speeds are exceeded. It emerges that North America is on average less windy than the United Kingdom and that light winds occur over a much larger proportion of its land surface. However, it is significant that at higher wind speeds the position is reversed. While only 4 per cent of the United Kingdom experiences average winds of $6\,\mathrm{m\,s^{-1}}$ or stronger the proportion of North America which is equally windy is 7·6 per cent. The North American wind regime exhibits greater variation than does that of the United Kingdom.

Figure 3.5 illustrates the windiness of North America in greater detail.

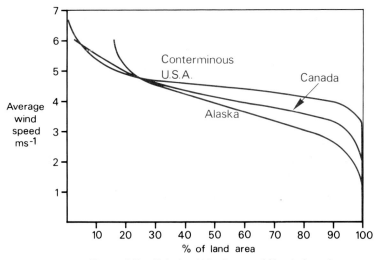

Figure 3.5   Relative Windiness of North America

It will be seen that over most of their surface area the conterminous United States, Canada, and Alaska rank in that order of decreasing windiness. Canada, however, has a much larger proportion of high wind speeds. The explanation of this fact is to be found in the extensive windy areas covering Keewatin, western Baffin Island, and the south and central portions of the arctic archipelago.

If good wind energy sites for 100,000 aerogenerators of the 91 m diameter GEC type could be found in North America their combined annual output would amount to 5550 PJ of electricity. This is nearly 52 per cent of the electrical energy delivered to consumers in the North American economy in 1979.

UNITED KINGDOM WIND VELOCITY DISTRIBUTION GRAPHS

These are shown in Figures 3.6 to 3.10 on pages 54–58.

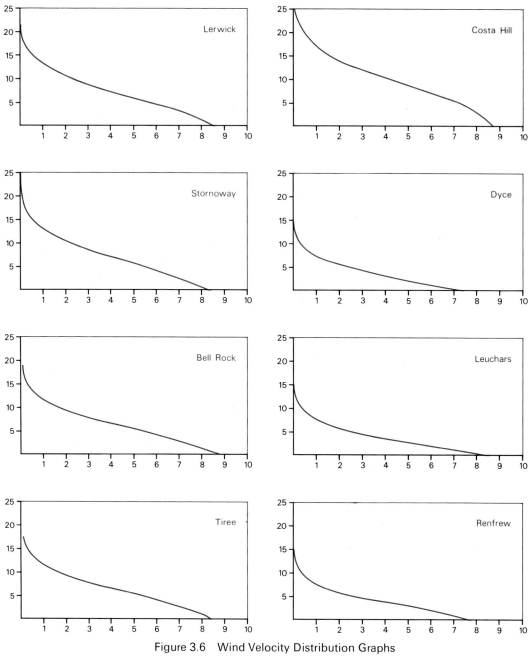

Figure 3.6 Wind Velocity Distribution Graphs

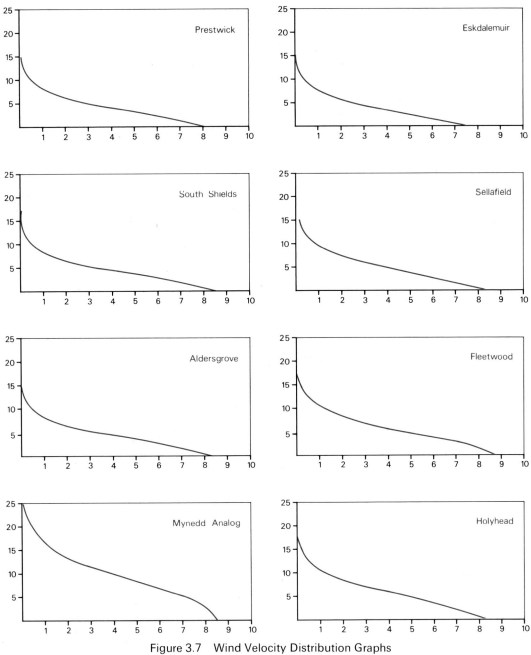

Figure 3.7 Wind Velocity Distribution Graphs

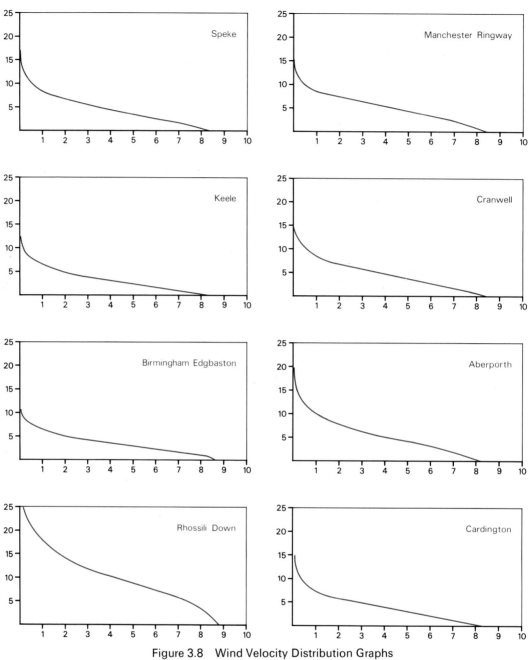

Figure 3.8 Wind Velocity Distribution Graphs

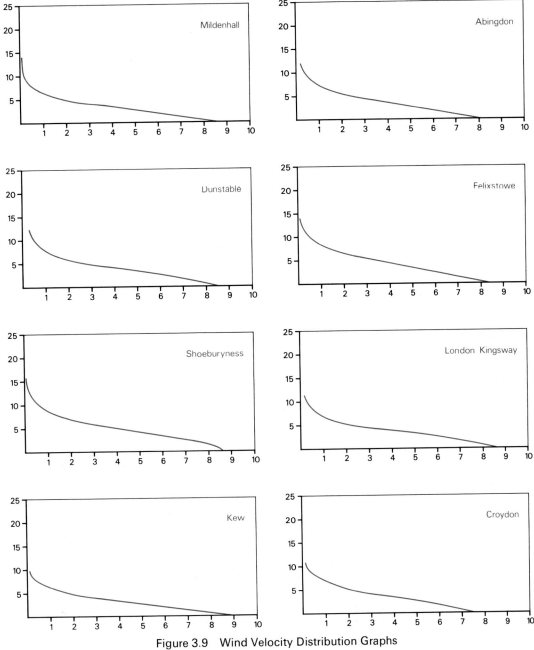

Figure 3.9 Wind Velocity Distribution Graphs

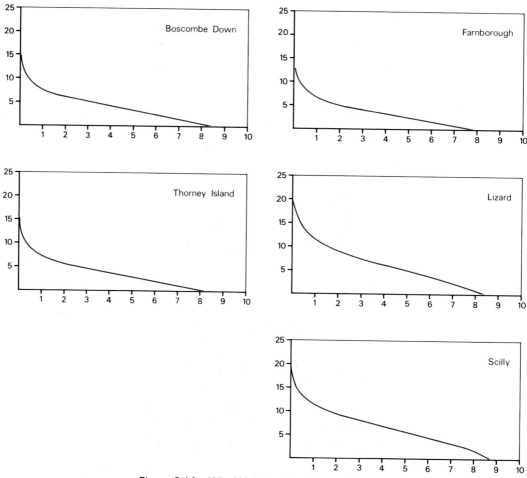

Figure 3.10 Wind Velocity Distribution Graphs

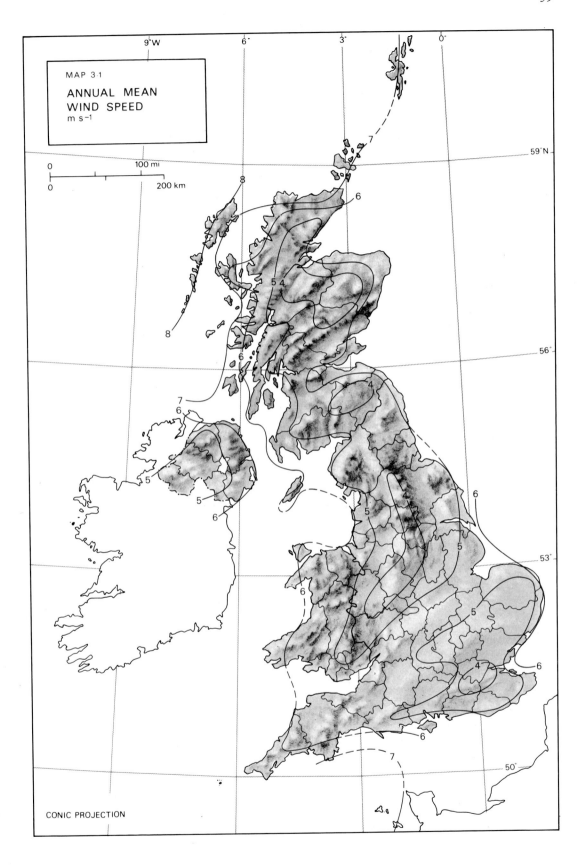

MAP 3.1 ANNUAL MEAN WIND SPEED m s⁻¹

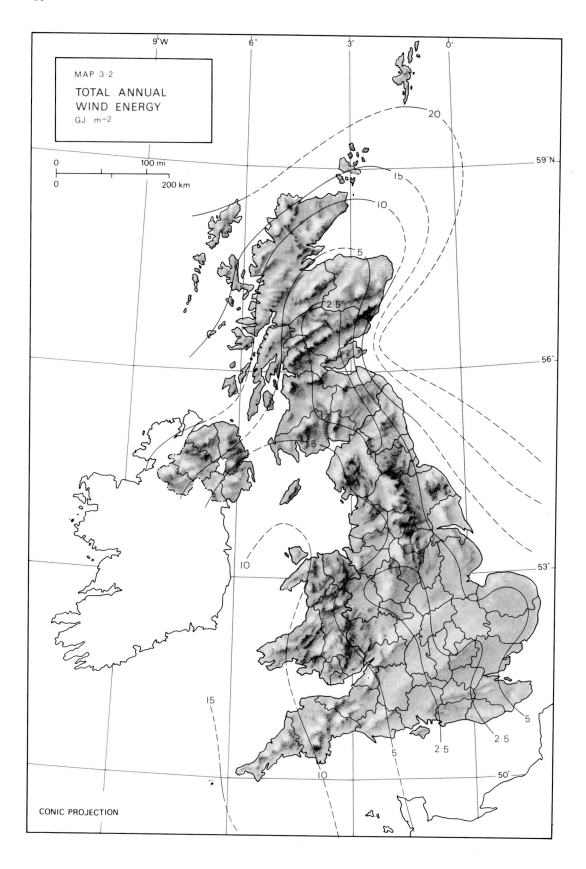

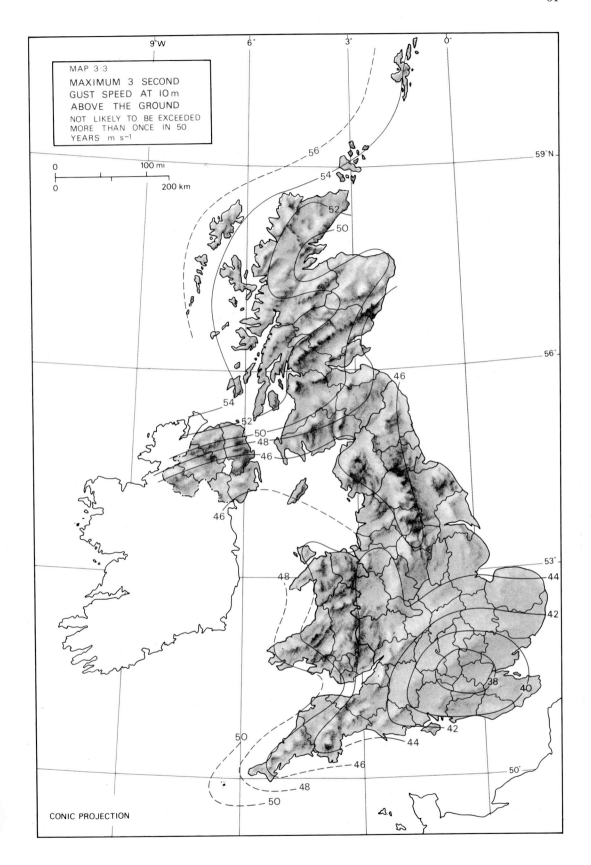

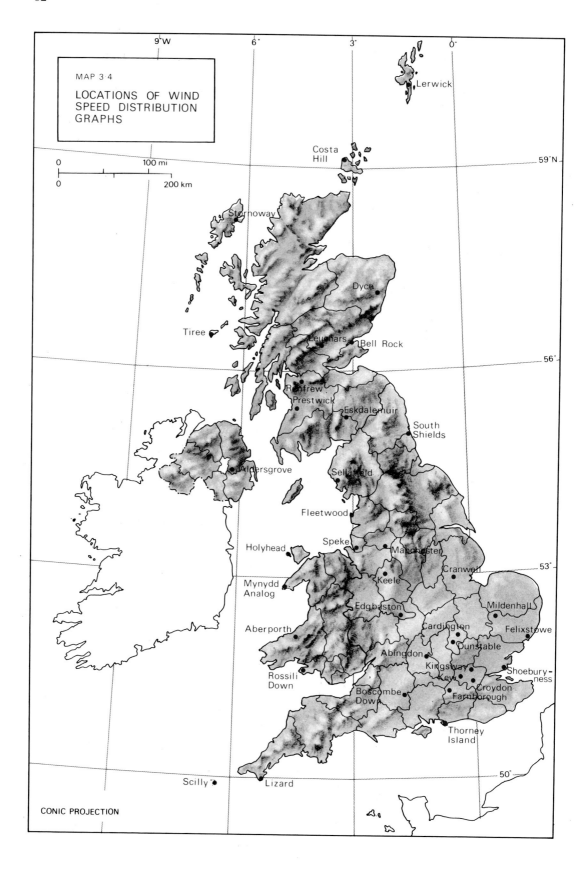

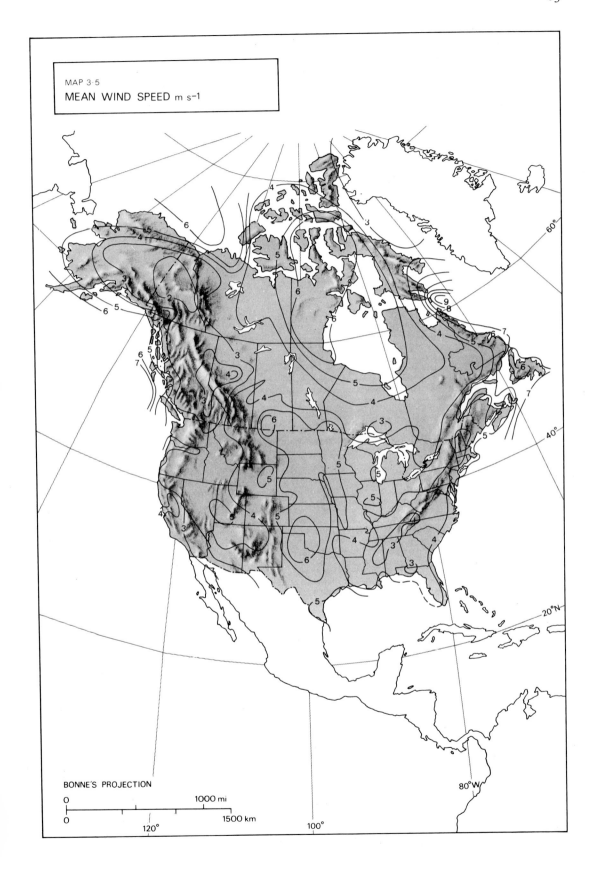

Chapter Four
Wave Energy

THE NATURE OF THE WAVES

When a body of water receives energy by friction from the motion of air across its surface the upper layers of the water will lose their equilibrium and be set in motion. In nature, winds blowing across the seas and oceans will transmit a part of their kinetic energy to the waters beneath and so produce the characteristic form of rhythmic fluid movement known as wave motion. A wave, as well as receiving energy from the wind, is engaged in a continuous process of absorbing energy from the next wave upwind and transmitting energy to its downwind neighbour. Thus it is that at sea trains of waves run with the wind.

A light wind of less than 1 m s^{-1} will raise small wavelets or ripples which fall away immediately the breeze dies. Above that wind speed, however, more stable gravity waves form and progress across the water with the wind. Although it can absorb only a proportion of the energy of the wind blowing overhead, water is so much heavier than air that it can store the energy it acquires in a very dense form. Gravity waves may therefore be considered as concentrators of wind energy, and for this reason they can become a valuable source of energy for man.

The surface of a wave in deep water takes the form of an inverted cycloid. This curve is the path traced by a point inside the rim of a circle rolled along a flat surface. Figure 4.1 illustrates the main features of a gravity wave.

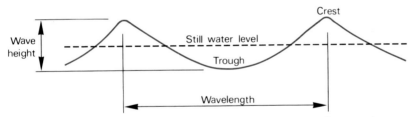

Figure 4.1 Features of a wave. (The time elapsing between the passage of successive crests or troughs is known as the wave period)

It is a common misconception that an advancing train of waves represents a body of water in forward motion. In fact a particle of water in a wave follows a circular path, advancing as the crest passes and retreating in the trough. Figure 4.2 greatly exaggerates, for the sake of clarity, the slight forward motion that takes place in each cycle. The advance is so small that a floating object, raised and lowered by a wave, is moved but little from its original position. It is as well that this should be, for the power of a moving body of water is so great that no ship could make headway against it, nor could any man-made structure long withstand a mass of water falling upon it at the speed of a travelling wave.

The size of the circular path of a water particle increases as the height of the wave increases and in all waves it reduces rapidly with depth. At the depth of half a wavelength the diameter of the circle is only 1/23rd of its size at the surface.

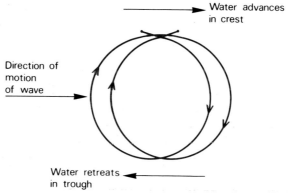

Figure 4.2 Motion of Water in a Wave

Figure 4.3 Water Motion within a Wave

The energy content of a wave similarly decreases with depth so that over 95 per cent of its energy is present in the top quarter wavelength of depth. Only under the most exceptional meteorological circumstances, therefore, is the water at a greater depth than 200 m affected by surface waves. Although moved by currents the deep waters of the ocean are forever undisturbed by the winds and storms that pass above them.

The speed with which a wave advances across the surface of the water can be found (Krummel, 1911) from the following formula:

$$C = (gL/2\pi)^{1/2} \text{ m s}^{-1}$$
where L = wavelength (m)
g = acceleration due to gravity (9.8 m s^{-2})

The wavelength and period of a wave are related by:

$$L = gT^2/2\pi \text{ m}$$
where T = period (s)

Figure 4.4 shows these relationships in graphical form. It will be observed that the speed of a wave is proportional to the square root of its wavelength. However, the time needed for a wave system to travel a given distance is twice that taken by an individual wave. This is because a leading wave is gradually overtaken by the wave following and, as it is absorbed, it transfers its energy to the overtaking wave. Every wave behaves in this way so that the complete system advances downwind at just half the speed of an individual wave.

THE QUANTIFICATION OF WAVE ENERGY

The size and character of the waves of the ocean depend upon a number of physical factors including wind speed, the length of time for which it has blown, the depth of the water, the shape of nearby land, and the fetch of the waves. Fetch is the distance over which the waves have built up uninterruptedly under the influence of the wind. The relationship between wind speed, wave height, and fetch is shown in Figure 4.5.

Only small waves form even in strong winds when the fetch is less than 10 km and for this reason only ocean waves are of interest for the purpose of energy production. However, the steep angle of the graphs between fetches of 10 and 100 km indicate a rapid build-up of wave height over moderate fetches. Little increase in height occurs, even in very strong winds, at fetches of more than 600 km while in moderate winds waves do not much increase beyond a fetch of 200 km.

It should be noted that Figure 4.5 gives the height which will be attained if the wind blows long enough for the waves to develop to their fullest extent. In the case of 25 m s^{-1} winds, as much as 39 hours is needed, as well as a fetch of 1000 km, to generate waves 14 m high. A system

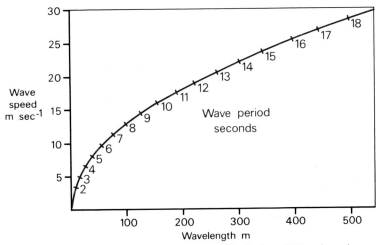

Figure 4.4 Wave Speed, Period, and Wavelength

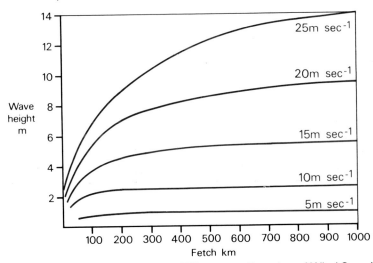

Figure 4.5 Size of Fully Developed Waves as a Function of Wind Speed and Fetch

of such large waves will therefore occur but rarely and only as the consequence of a great storm blowing across a broad ocean.

Ocean wave records use a height measurement known as significant wave height, which is defined as the average height of the highest third of the waves. Figure 4.6 gives the frequency with which waves of various significant height occur at ocean weather ship India stationed west of the Hebrides at 56° 31′ N 20° 10′ W.

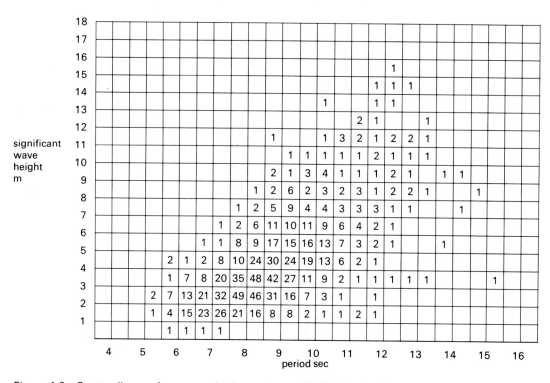

Figure 4.6 Scatter diagram for one year's observation at OWS India showing the number of occurrences in parts per thousand of waves of a particular height and period

The energy of a wave can be calculated to a sufficient degree of accuracy by considering the surface of the water to be sinusoidal. It can then be shown (Salter, 1974) that the power of a wave can be found from the following formula. It will be noted that the power of a wave is proportional to the square of its height.

$$P = (W\rho g^2 T H^2)/64\pi \text{ W}$$

where W = width of wave front (m)
ρ = density of seawater ($1.03 \times 10^3 \text{ kg m}^{-3}$)
g = acceleration due to gravity (9.81 m s^{-2})
T = wave period (s)
H = significant wave height (m)

When this expression is applied to each cell of the scatter diagram in Figure 4.6 and the results are averaged a figure of 77 kW m^{-1} of wave frontage is obtained at OWS India. The quantity of energy transported across a front of unit length is obtained by multiplying the power density by the number of seconds in the time interval concerned. Appropriate factors are given in Appendix 1.

THE WAVE ENERGY RESOURCE

Measurements of wave phenomena have been made in the waters around the British Isles for many years. The purpose of the work has usually been military, navigational, or for the construction of coastal protection works, but in recent years some measuring devices have been deployed for academic and commercial research into waves as a source of useful energy. These devices are too sparsely distributed, however, to allow a synoptic picture of the United Kingdom wave energy regime to be obtained by means of them. Furthermore, the results from some recording stations are not generally available for reasons of commercial confidentiality.

The most commonly used wave measuring device is the design known as a Waverider. It consists of a floating steel sphere 0·7 m in diameter moored to the sea bed and containing apparatus to measure wave height, period and direction as well as water temperature and salinity. Although the device can transmit data to the shore by radio its range is limited and all its equipment requires regular maintenance and adjustment. A measurement programme calling for the deployment of many Waveriders around the coasts would, though most desirable, be a very expensive undertaking. Resort must therefore be made to calculation and mathematical simulation of the ocean wave regime.

In 1977 there was published under the auspices of the Meteorological Office a generalized mathematical model capable of deriving ocean wave fields from meteorological data. This depth dependent model takes into account energy loss and spreading mechanisms such as bottom friction, wave breaking and refraction round obstacles. It does not, however, allow for the effects of geography and the shape of nearby coastlines. Additional work to apply the model to the coastal waters of the United Kingdom has been done by the Energy Technology Support Unit at Harwell and the resulting adapted model has been used to calculate the size of the wave energy resource at 71 points off the north and west coast of the British Isles (Winter, 1980). The model results show a good correlation with the available Waverider data and they have been used to prepare Map 4.1.

THE MAPS

If a wave encounters a wave power machine that is perfectly efficient it will surrender all its energy to the machine and leave the surface of the water to leeward quite calm and flat. A train of waves will give up their energy to a machine continuously and will therefore produce power, which is energy active over time. The isopleths on Map 4.1 have been drawn to connect points of equal wave power per metre of wave frontage. The quantities of power are those that would be delivered to a machine of 100 per cent efficiency.

The map reveals a number of features of the wave energy situation in Britain. Energy densities are much higher on western coasts than in the English Channel and the North Sea. This is to be explained by the long fetch of waves arriving from the west across the Atlantic Ocean. Energy densities encountered off the Hebrides are slightly higher than those in southwest England because of the greater wind speeds experienced at higher latitudes.

Energy densities increase with distance from the shore for two reasons. In the first place friction with the sea bottom deprives waves in shallow water of some of their energy. Even in water 50 m deep a wave will be attenuated by bottom friction. Secondly, offshore locations are exposed to the arrival of waves from all points of the compass and therefore receive a larger total of energy than coastal situations which are sheltered from one or more directions.

In Map 4.2 the power density data of Map 4.1 have been converted into quantities of energy available each year per metre of wave frontage. The geographical pattern of energy availability naturally corresponds to the distribution of wave power density. An assessment of the amount of energy available from waves may be gained by comparing Map 4.2 with Map 2.15 showing annual totals of solar radiation. In the case of the island of South Uist in the Outer Hebrides one metre width of wave frontage delivers 500 times as much energy each year as could be gathered from the sunlight falling upon one square metre of a flat horizontal surface. The respective quantities are $1.5\,\text{TJ}\,\text{m}^{-1}$ and $3.0\,\text{MJ}\,\text{m}^{-2}$. These figures are eloquent of the ability of ocean waves to concentrate wind energy and therefore, in effect, to concentrate the energy of the sun.

THE SIZE OF THE WAVE ENERGY RESOURCE

The amount of energy present in ocean waves is enormous. Observations made for many years at OWS India show that waves in the open Atlantic could deliver $3.3\,\text{TJ}\,\text{m}^{-1}$ of energy a year. Since the average semi-detached house consumes about 350 GJ a year in space heating, cooking, and providing hot water it is evident that a metre length of wave machine moored in the position of OWS India could, at 100 per cent efficiency, supply all the energy needed by about nine houses. However, there is little to be gained by taking this comparison further, since OWS India is too remote and in too deep water to be a practical wave energy site.

Wave energy machines are suitable for generating electricity, and a better estimate of the resource can be obtained by considering its potential in relation to the amount of electrical energy consumed by the United Kingdom economy. This, in 1979, totalled 848 PJ. A wave energy system will suffer mechanical losses in generation and transmission losses both to the land and in subsequent distribution ashore to the consumer. The overall efficiency of a fully developed wave energy machine is likely to be about 30 per cent. A line of wave energy conversion machines moored along the $2\,\text{TJ}\,\text{m}^{-1}$ isopleth to the west of the Hebrides and the Scilly Isles would need to be, at this efficiency, about 1400 km long if all the nation's electricity were to be obtained from this source. Map 4.3 has been marked with a line representing 1000 km of machines deployed on the $2\,\text{TJ}\,\text{m}^{-1}$ isopleth. They could deliver to the consumer 600 PJ of electricity annually, some 70 per cent of the 1979 demand for this form of energy.

WAVE ENERGY IN NORTH AMERICA

The wave energy resource represents a gap in the otherwise remarkably complete body of published information on renewable energy regimes in Canada and the United States. Very few measurements of wave energy at sea have been carried out. Furthermore, the wave climate of either the east or west coast has yet to be modelled after the manner of the Harwell study of wave conditions in British waters. Some indication of the size of the North American resource may be gained, however, from information gathered during the Seasat mission of 1978 (Chelton *et al.*, 1981).

Three short wave radars and two passive radiometers were carried aboard the Seasat spacecraft from 28 June until its breakdown on 10 October. The instruments were used to measure and relay to Earth data on atmospheric liquid water, water vapour, surface wind speed, and ocean wave heights. The information gathered has enabled these phenomena to be mapped on a worldwide basis for the first time. Maps drawn from Seasat data delineate significant wave heights, at least for the summer months of 1978, on both coasts of North America. The results are given on Map 4.4.

It will be observed that the significant wave height on the east coast of the North American continent does not exceed one metre except off Newfoundland and Labrador. More boisterous conditions prevail on the west coast, however, which lies at the downwind end of the North Pacific pattern of weather circulation. Here wave conditions are very similar to those found on the western side of the British Isles, where waves also arrive from across an ocean. An approximate idea of the North American wave energy resource can be obtained by supposing that the same energy potential exists in Britain and North America at locations where significant wave heights are the same.

On this assumption wave energy machines deployed along the line drawn on Map 4.5 some 50 to 100 km off the west coast of Canada and the United States would receive $2\,TJ\,m^{-1}$ of wave energy per year. The line is 3000 km long. By the reckoning applied to the United Kingdom wave energy resource these machines would be capable of delivering 1800 PJ of electricity a year to the North American economy.

The east coast of the United States, like the North Sea side of the British Isles, cannot be as productive of wave power as western districts. A line of wave energy devices moored along the line drawn on Map 4.5 some 100 km off the east coast would probably receive, because of the relatively calm wave climate, about $1\,TJ\,m^{-1}$ a year. A front of 2000 km of machines deployed here would be able to deliver 600 TJ of electricity annually. The wave energy resource of the North American continent may therefore be put at 2400 PJ of electricity annually obtained from 5000 km of wave energy machines.

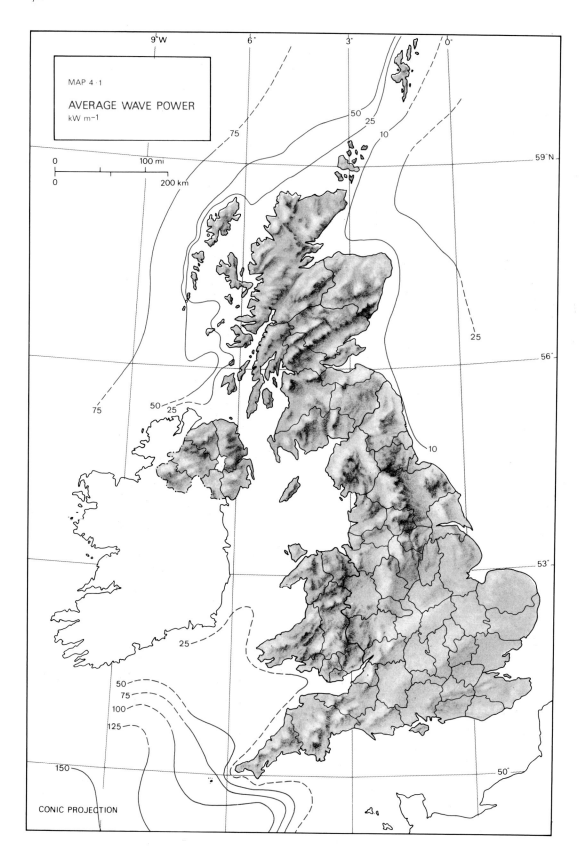

MAP 4·1

AVERAGE WAVE POWER
kW m⁻¹

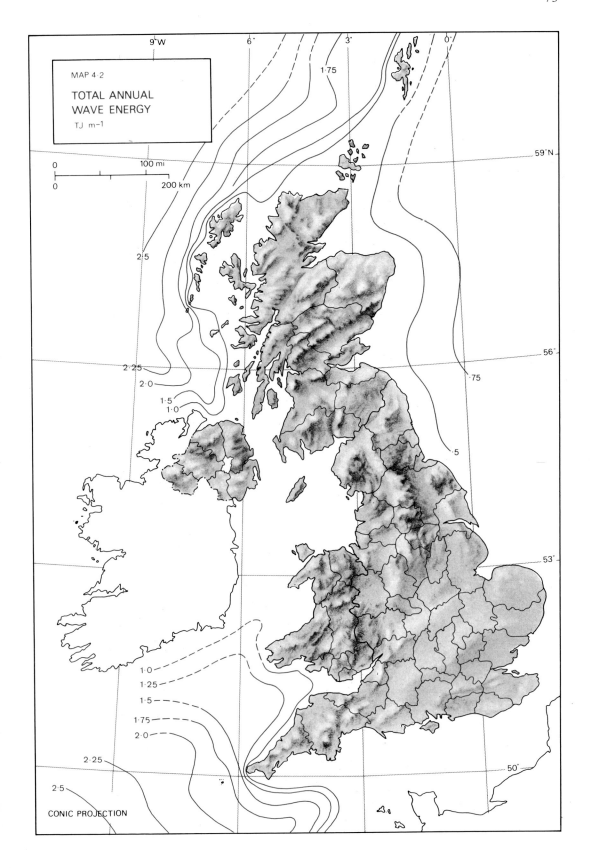

MAP 4.2

TOTAL ANNUAL WAVE ENERGY

TJ m⁻¹

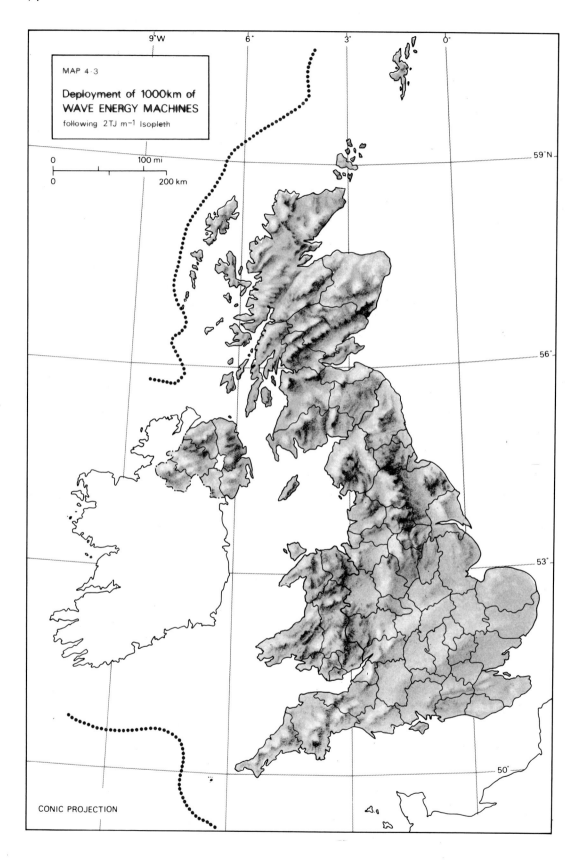

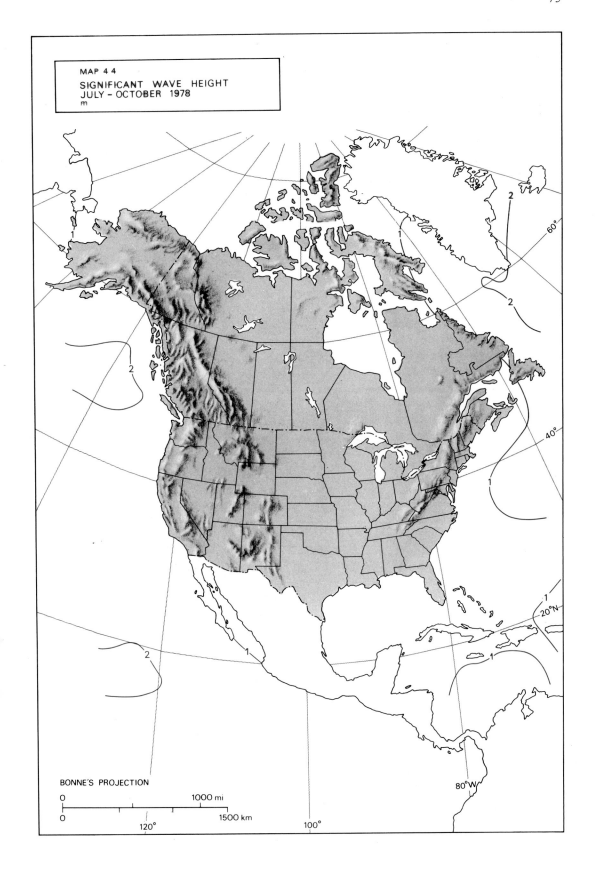

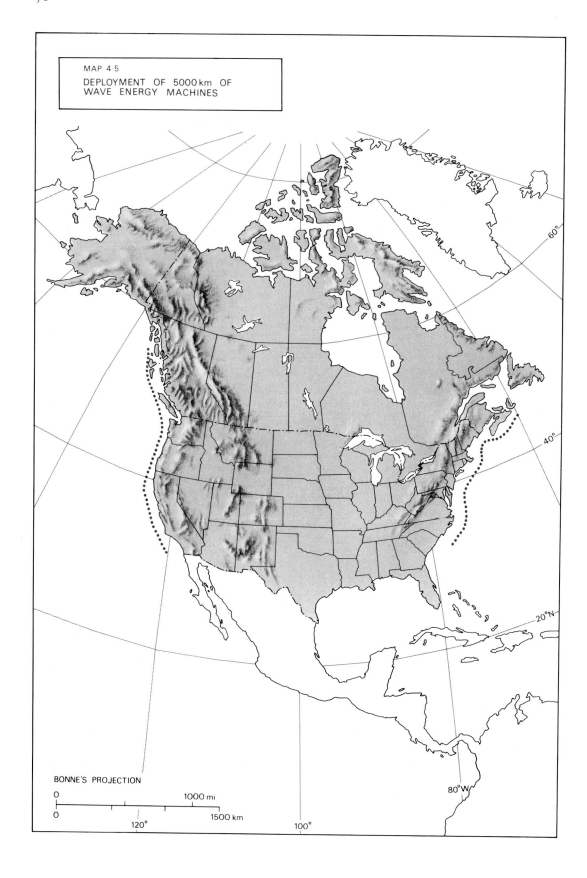

Chapter Five
The Thermal Energy of the Oceans

More than two-thirds of the surface of the globe consists of water. Consequently, about twice as much of the Sun's incident radiation is intercepted by water as is received by the world's land areas. Some of the sunlight falling onto the surface of the oceans and seas is not absorbed but is reflected back into space. The shorter wavelengths are reflected by the water more strongly, with the result that the Earth when seen from space appears to be a blue planet. In Figure 5.1 the proportion of incident radiation reflected is plotted against the angle of incidence of the Sun's rays. The fact that reflection of direct sunlight increases rapidly when the incident angle falls below about 30 degrees has important consequences for the geographical distribution of ocean thermal energy.

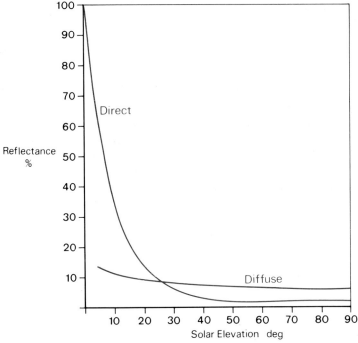

Figure 5.1   Relationship between Reflectance and Angle of Incidence

Because seawater is translucent a large proportion of the arriving solar radiation penetrates beneath the surface, but it does not reach to any great depth. Figure 5.2 shows that light reaches down to only about 65 m in average oceananic water and that even the slightly more penetrating blue part of the spectrum is completely absorbed at about 90 m. Therefore it is only in the uppermost layers of the ocean that the energy of the Sun's radiation is absorbed and stored in the water as heat. At a depth of about 100 m the last remnants of the Sun's rays are extinguished, leaving the ocean depths in perpetual darkness and cold.

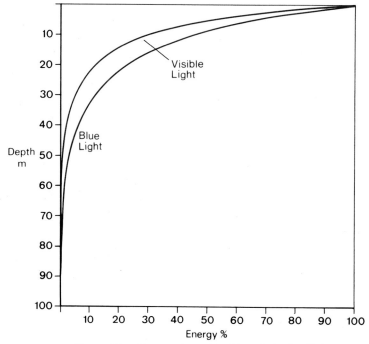

Figure 5.2   Penetration of Incident Solar Radiation

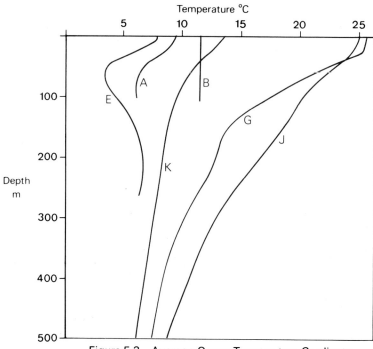

Figure 5.3   Average Ocean Temperature Gradients

In effect, the top 50 m of the ocean functions as the largest solar collector on Earth. The warmed waters of this absorbing layer lie over the colder and denser waters of the depths to form a stable temperature stratification. Although tidal streams, ocean currents, and turbulence from surface waves can disturb the waters and mix the layers to some degree, it remains generally true that, except at high latitudes, the temperature of the ocean varies inversely with depth.

At latitudes above 50 degrees the Sun reaches an altitude of no more than 17 degrees in winter, while at 60 degrees latitude the maximum mid-winter solar elevation is only 7 degrees. Figure 5.1 shows that the proportion of the incident radiation reflected at mid-winter noon from the surface of the sea at these latitudes 17 per cent and 47 per cent respectively. In addition, the intensity of solar radiation at high latitudes is diminished by the fact that the path of the Sun's rays through the Earth's atmosphere is long when the solar elevation is low. The high proportion of the attenuated insolation that is reflected in these regions means that the seas and oceans of high latitudes can obtain only a small amount of heat from the Sun. Consequently, in polar regions the normal temperature gradient in the ocean is reversed, with sea ice floating upon unfrozen deeper water. In practice only tropical seas receive and retain enough solar energy to produce a thermal gradient large enough to drive an ocean thermal conversion machine.

Figure 5.3 shows the rate of fall of temperature with depth at six locations in the seas around the United Kingdom and North America. The position of points A and B are marked on Map 5.1 of the British Isles while Map 5.2 shows points E, G, J, and K in North American waters. The values plotted are annual average temperatures at the various depths. It will be noted that only at location G off the coast of Florida and location J in the Gulf of Mexico does the gradient exceed 10 °C in the upper 250 m of water. At location B in the English Channel, where strong tidal streams keep the shallow waters well mixed, the average temperature gradient is nil.

CONVERSION OF OCEAN THERMAL ENERGY

The chief requirements for the functioning of a heat engine are a heat source and a heat sink. High efficiencies call for a large temperature difference between the hot and cold sides of a heat engine, but this is by no means an indispensable requirement of their operation. The small temperature differences existing between the layers of the ocean can be used to drive suitably constructed machines. A number of large floating ocean thermal energy conversion, or OTEC, devices for generating electricity by means of low temperature turbines have been designed, and at the time of writing the construction of prototype machines to test the usefulness of this energy resource has begun.

It can be shown (Mangarella and Heronemus, 1979) that the electrical power able to be extracted from the thermal gradient of a body of seawater can be found from the following formula:

$$P = 0.12 \, dT_a^2 / (T_h + 273) \, \mathrm{MW \, m^{-3} \, s^{-1}}$$

where dT_a = difference in temperature between hot and cold sources (°C)
T_h = hot side temperature of the engine's working fluid (°C)

This equation assumes an overall Carnot efficiency of 60 per cent and gives the power production in terms of the rate of flow of hot seawater through the machine in $\mathrm{m^3 \, s^{-1}}$.

If a value of 20 is assumed for dT_a and 24 °C for T_h it can be calculated that an OTEC device will produce only 0.16 MW of electrical power when the flow of water is $1 \, \mathrm{m^3 \, s^{-1}}$. Very large quantities of seawater must therefore be moved if useful quantities of energy are to be produced. At these operating temperatures a flow rate of $619 \, \mathrm{m^3 \, s^{-1}}$, calling for very large capacity pumps and ductways, would be required to produce electricity at the rate of 100 MW.

It will be noted that the output of an OTEC machine varies directly with the square of the temperature difference between the heat source and the cold sink. It is for this reason that oceanic regions possessing a large thermal gradient are needed for the production of power from the ocean thermal resource.

THE MAPS

The British Isles rest upon a large northwestward extension of the European continental shelf and in consequence the coastal waters of the United Kingdom are shallow. Most of the Irish Sea, the eastern part of the English Channel and the southern half of the North Sea are less than 50 m deep. The entire group of islands is enclosed within the 200 m submarine contour and the 1000 m contour line is at its nearest some 110 km beyond the Outer Hebrides. Large temperature differences cannot exist within these waters because any temperature gradient will be short and also because British coastal waters experience strong tidal streams. The regular ebb and flow of the tides keeps the water well mixed and inhibits the establishment of a thermal gradient.

Figure 5.4 shows the difference between the temperature at the surface and at 50 m or 100 m throughout the year at the four locations marked on Map 5.1. At location A a gradient of about 7 °C occurs in August diminishing to zero in winter. In the English Channel at location B the water is isothermal at all times of the year while at C and D small positive summer gradients become negative in the winter months. It must be concluded that conditions necessary for the operation of OTEC devices do not exist in the United Kingdom.

In North America, however, the thermal state of the water is very different. Map 5.2 charts the annual average difference between the temperatures of the surface and that of water at a depth of 1000 m. The gradients are small on both the Atlantic and Pacific coasts of Canada but further south at locations K, L, and M off the west coast of the United States differences of as much as 11 °C are found. West coast gradients are fairly uniform throughout the year as is shown in Figure 5.5.

On the east coast of the United States a small uniform temperature difference exists throughout the year at location F off Cape Hatteras. At location G a gradient of nearly 15 °C is found all the year round between the surface and a depth of only 500 m. This location, where the Florida current flows north through the relatively narrow strait between Florida and the Bermuda Islands, is the most favoured OTEC site in North America. In the Gulf of Mexico large gradients are to be found for most of the year, although at both locations H and J the temperature difference to 500 m falls for a time in winter to only about 10 °C. Map 5.2 shows that in North America it is the southeastern states of the United States that are best placed to make use of the ocean thermal energy resource. The countries on the eastern shore of Central America and the islands of the Caribbean Sea, where gradients to 1000 m exceed 20 °C, are in an even better position to obtain energy from this source.

THE SIZE OF THE RESOURCE

Full scale OTEC machines are so large, and extract heat from so great a volume of water, that their pumping operations are capable of reducing or eliminating a temperature gradient occurring naturally in an otherwise undisturbed body of seawater. A stagnant mass of water of restricted size would ultimately be made isothermal by the mixing together of cold water from the depths with the warm surface layers. Sites suitable for the deployment of OTEC machines therefore require not only a large thermal gradient but also a steady horizontal current of water to maintain the stability of the gradient. In effect, the current is needed to deliver to an OTEC site the heat energy to be extracted by the machines.

At location J in the Gulf of Mexico the surface current flows in various directions at an average speed of about $0\cdot2\,\mathrm{m\,s^{-1}}$. The quantity of water transported in the upper 50 m across a front 50 km wide is therefore $5\times10^5\,\mathrm{m^3\,s^{-1}}$. A line of OTEC machines may be supposed to be deployed across a front of this length and to be capable of intercepting and utilizing for the generation of electricity one-fifth of the flow of warm water. If T_a is taken to be 18 and T_h to be 21 °C then the rate of electricity production of such an array of machines would be 13 GW, amounting to 420 PJ of electrical energy in the year. If each OTEC device was rated at 100 MW the array would consist of 130 machines.

Thermal conditions at location H, off the mouth of the Mississippi River, are the same as at location J and a similar 50 km long array of 130 OTEC machines would produce a further 420 PJ of electricity a year at this site.

The current flowing north through the Florida Strait is much faster than those found in the

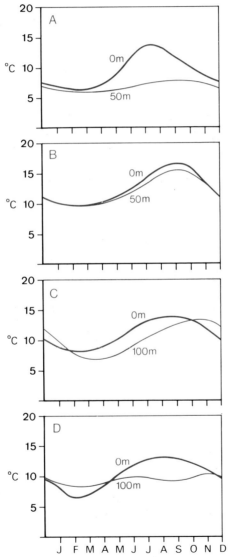

Figure 5.4 Seasonal Variation of Temperature in British Waters

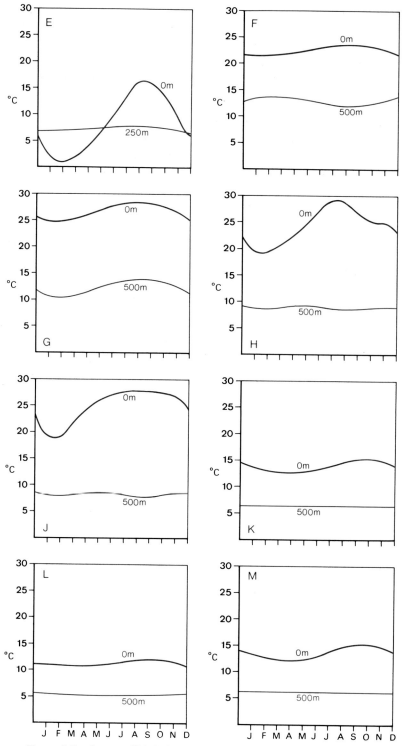

Figure 5.5 Seasonal Variation of Temperature in North American Waters

northern part of the Gulf of Mexico. The average speed of the fastest moving part of the Florida Current, between Miami and Bimini, is $1\cdot3\,\mathrm{m\,s^{-1}}$. A line of OTEC devices intercepting this flow of water across a 10-km front would in the upper 50m be presented with $6\cdot5 \times 10^5\,\mathrm{m^3\,s^{-1}}$ of warm water. A calculation using the same temperatures as at locations J and H shows that 172 machines of 100 MW each would be able to produce 542 PJ of electrical energy a year at location G on Map 5.2.

The energy available from locations G, H, and J is summarized in Figure 5.6. The table compares OTEC electricity production potential with the consumption statistics of the markets to which it could most readily be transmitted. Consumption figures are those published for 1979 and the table shows that 59 per cent of the electricity sold in the three southeastern regions of the United States could be provided by 432 OTEC machines working from the ocean thermal resource.

Region	Electricity consumption 1979 PJ	OTEC site	Number of 100 MW machines	Electricity production by OTEC PJ	Less 10% transmission losses PJ	Percentage of 1979 consumption from OTEC
Southeast	1108	G	172	542	488	44
Delta	285	H	130	420	378	133
Plains South	708	J	130	420	378	53
Totals	2101		432	1382	1244	59

Figure 5.6 OTEC Electricity Production in North America

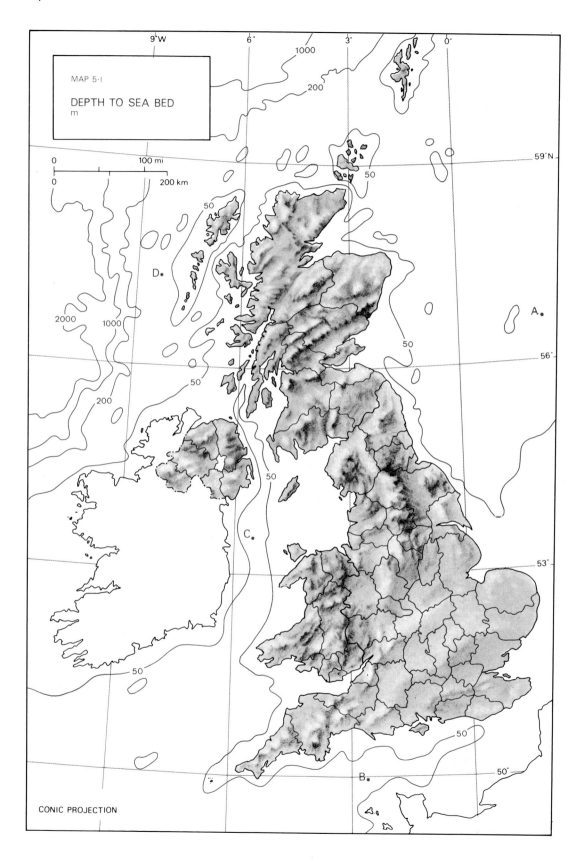

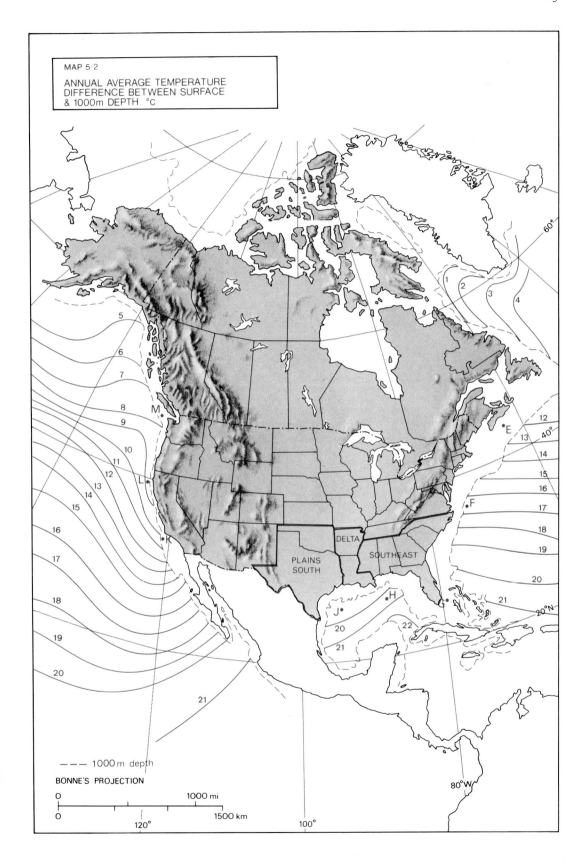

Chapter Six
River Energy

THE NATURE OF RIVER ENERGY

Nearly a quarter of the energy arriving at the Earth from the Sun is used in the processes of evaporation, convection, precipitation, and surface runoff of water. The operation of this vast solar-powered heat engine, known as the hydrological cycle, is responsible for the formation of clouds, fog, rain, hail, and snow, and for the creation of snow fields, glaciers, lakes, and rivers. This chapter is concerned with harnessing the energy imparted to flowing river water by the perpetual revolution of the hydrological cycle.

The size and power of a river is dependent upon the extent of its catchment basin, the amount of precipitation received by the basin, and the distance fallen by the waters of the river on their journey down to the ocean. The potential energy possessed by water placed in a river bed by the hydrological cycle is, in nature, dissipated by friction into low grade heat. Some of this energy can be trapped, however, and by controlling the water's pattern of flow and method of descent some of its energy can be harnessed for the purposes of mankind.

A river flowing over hard impervious rock will exhibit large alterations in its rate of flow in response to changes in rainfall and snowmelt, while a porous subsoil will absorb water and, acting as a capacitor, smooth the peaks and troughs of the water's flow. The higher and steeper parts of a river channel are usually of the first type with the flow becoming more even as it crosses the valley floor near to the sea. But tempestuous or sluggish, the energy potential of a river is a function of its rate of flow, its slope, and its length.

THE QUANTIFICATION OF RIVER ENERGY

The gross amount of potential energy dissipated each year by a flowing river can be found from the following formula:

$$E = 0.155 \ LSQ \ TJ$$

where L = river length (km)
S = slope (m km^{-1})
Q = average rate of flow (m^3 s^{-1})

Some of a river's energy can be harnessed by a water wheel or turbine that is rotated by the water's unregulated flow. Most pre-industrial water mills were of this 'run of the river' type with the result that their output varied with the state of the river. In times of spate water would run to waste for lack of capacity in the wheel, while a drought could bring operation to a halt.

In modern practice it is usual to avoid this difficulty by constructing an energy store in the form of a reservoir behind a dam or barrage built across the river valley. The discharge of water from the reservoir can then be controlled so as to regulate power production, smooth output, and match output to demand. Barrages are comparatively easy to build at places where a river runs through a gorge, for there a small structure can be made to impound a large volume of water. A river flowing through a wide shallow valley will, however, require a large and expensive engineering effort if it is to be formed into a reservoir.

River power reservoirs are often very extensive, and in settled country a proposal to flood a valley can be in conflict with other uses for the land. Even in remote country a barrage and reservoir will alter the natural environment profoundly and may constitute an unwelcome visual intrusion. The amount of energy that can in practice be obtained from a river is therefore a function of the geomorphology of its catchment area and of the other uses in competition for the basin. It follows that the physically exploitable energy potential of a river can only be discovered by means of a detailed field survey. Two superficially similar basins can, when surveyed, be found to have quite different energy possibilities and no general rule relating gross and physical energy potential is available.

THE SIZE OF THE RESOURCE

The thirty largest rivers in the United Kingdom are listed in Figure 6.1 in order of their average rate of flow. When, however, the gross energy potentials of these rivers are calculated a different ranking emerges. The reordering results from the fact that gross energy potential is a function of slope and river length as well as flow rate. The 25 most energetic rivers in Britain are listed in

River and gauging station		Average Flow $m^3\ s^{-1}$	Slope $m\ km^{-1}$	Length km
TAY	Ballathie	152.21	10.78	109.88
BANN	Monanagher Weir	105.00*	6.59	101.37
TRENT	Colwick	82.21	4.16	148.99
TWEED	Norham	73.85	5.95	140.34
NESS	Ness Castle Farm	73.14	10.62	104.61
WYE	Cadora	71.41	3.31	224.59
THAMES	Teddington	64.40	1.36	238.75
SPEY	Boat O'Brig	62.80	8.39	151.04
SEVERN	Bewdley	62.70	3.93	206.32
BEAULY	Erchless	44.44	18.41	61.69
TYNE	Bywell	43.45	9.89	88.66
CONON	Moy Bridge	40.82	20.93	52.55
OUSE	Skelton Rlwy Bridge	40.45	6.33	111.91
LOCHY	Fort William	40.15*	16.65	79.10
DEE	Chester Weir	40.00*	6.62	132.85
LEVEN	Linnbrane	39.75	18.33	61.43
TYWY	Ty Castle Farm	38.34	9.62	81.52
CLYDE	Daldowie	36.90	6.43	112.82
AIRE	Beal Weir	36.89	5.33	110.25
DEE	Woodend	35.70	10.69	116.27
LUNE	Halton	34.65	12.14	60.32
RIBBLE	Salmesbury	31.72	7.15	93.95
EDEN	Warwick Bridge	31.02	9.18	101.53
TEIFI	Glan Teifi	28.15	5.98	98.35
USK	Chain Bridge	27.37	9.84	87.68
EWE	Poolewe	24.75	18.97	52.92
NITH	Friars Carse	24.47	9.76	72.22
DERWENT	Camerton	23.84	17.04	54.77
TAMAR	Gunnislake	23.29	8.47	67.50
ESK	Netherby	22.78	11.11	61.02

*estimated

Figure 6.1 Largest British Rivers

Figure 6.2 and illustrated on Map 6.1. It will be noted that the river energy resource is located mainly in the northern and western parts of the country. Extensive areas of elevated country, resulting in steeply-sloping river profiles, together with high rainfall levels are responsible for the geographical distribution of river energy in the United Kingdom. The combined gross potential of these 25 rivers amounts to 184 PJ annually and would, if it could be harnessed, make a large contribution to the energy requirements of the British economy. But in practice only a small proportion of this total can be made available.

In 1943 a thorough survey was made of the hydro-electric generating potential of the Scottish Highlands and the results were published a year later (Lawrie et al., 1944) in Edinburgh. This work identified 102 hydro-power sites including 29 sites lying on the large Highland rivers listed in Figure 6.2. These nine large rivers, together with the potential of the Clyde and the Tweed, could furnish 9·8 PJ of electrical energy annually. The potential of all 102 sites enumerated in the 1944 report plus that of the Clyde and Tweed, totals 22·9 PJ.

For reasons of scale public electricity generating utilities are unable, except in special circumstances, to build and operate small power stations. Sites with a potential power output of less than 0·5 MW were therefore excluded from the survey of Scottish rivers. However, a considerable energy resource exists in the flow of smaller rivers and streams. Small hydro-power plants,

River	Gross potential PJ	Physical potential PJ
Tay	27·946	3·096*
Ness	12·595	2·232*
Spey	12·335	0·432*
Bann	10·872	0·217
Tweed	9·558	0·191
Wye	8·228	0·114†
Lochy	8·196	1·127*
Trent	7·898	0·158
Beauly	7·823	0·450*
Conon	6·959	1·314*
Leven	6·938	0·180*
Dee (Woodend)	6·878	0·324*
Severn	6·797	0·136†
Tyne	5·905	0·118
Dee (Chester)	5·453	0·163†
Tywy	4·660	0·100†
Eden	4·481	0·090
Ouse	4·441	0·089
Clyde	4·149	0·083
Lune	3·933	0·079
Ewe	3·852	0·396*
Usk	3·660	0·073†
Derwent	3·449	0·069
Aire	3·360	0·067
Ribble	3·303	0·066
United Kingdom	183·647	11·364

*Lawrie
†Wilson

Figure 6.2 Most Energetic British Rivers

while they are not suitable for connection to the national electricity distribution grid, can make a useful contribution to satisfying local demand. A detailed survey has recently been made of the rivers of Wales (Wilson et al., 1980) in order to discover their ability to produce energy from small hydro-electric plants. The survey identified 565 sites each able on a realistic assessment to produce 25 kW or more of power. The sites lying on the five principal rivers of Wales listed in Figure 6.2 could contribute about 0·5 PJ each year, a quantity of energy which, though slight nationally, would be useful if applied to meeting local demand. The annual energy potential of all 565 sites surveyed in Wales is about 1·0 PJ.

The energy potential of the rivers of England and Northern Ireland awaits survey and is not at present known. The physical potential of these rivers has, in the absence of observed data, been assessed at 2 per cent of their gross potential. The physical potential of all 25 rivers listed in Figure 6.2 then amounts to 11·364 PJ, only 1·3 per cent of the annual electricity consumption of the United Kingdom economy. It is clear that from a national point of view the rivers of Britain can be no more than a minor source of energy. However, in Scotland, and to a lesser extent in Wales, rivers could make a significant contribution to meeting local need. Figure 6.3 shows that small generating plants built on the rivers of Wales could furnish just over 2 per cent of the electricity at present consumed in the Principality, while nearly a quarter of Scottish electricity demand could be met from the river energy resource.

	Electricity consumption 1979 PJ	River energy potential PJ	Percentage
Scotland	96·76	22·86	23·6
Wales	51·43	1·08	2·1

Figure 6.3   River Energy in Scotland and Wales

RIVER ENERGY IN NORTH AMERICA

The land area of the United States, Canada, and Alaska together is 1836×10^6 ha, some 76 times that of the United Kingdom. Many of the river catchments of the continent are very large, and in the case of the Mississippi–Missouri river basin contains the longest river system in the world. The western part of North America is occupied by one of the world's highest and most extensive mountain chains, while the summits of the Appalachian Mountains and the mountains of Baffin Island, both small in comparison with the Rockies, rise higher than any British mountain. The North American climate is generally drier than the British but sufficient rain falls across the continent to fill the channels of many mighty rivers. The combined effect of these meteorological and topographical features is to produce a river energy resource far larger than that existing in the United Kingdom. In fact it is nearly 300 times as large.

Figure 6.4 lists the 30 largest rivers of North America in order of their rates of flow. The statistics on slope, length, and flow rate have been used to calculate their gross power potential and the results for the 25 most energetic rivers are shown in Figure 6.5. The courses of these rivers, above their gauging stations, are drawn on Map 6.2. Were it possible to exploit all their power, these rivers could supply the entire demand for electricity of the North American economy in 1979. Only a part of the resource is, however, available for use.

Surveys have shown that only the energy listed in the second column of Figure 6.5 could in practice be harnessed. The physical potential of the 25 rivers, 3322 PJ annually, is 31 per cent of their gross potential. It should be noted that the physical potential of the Churchill River and of La Grande Riviere in Eastern Canada exceed their gross potentials because large engineering works have diverted water from neighbouring rivers into their catchments.

The relative importance of river energy in North America is shown in Figure 6.6. Here the physical potential for electricity generation of the 25 rivers is compared with the 1979 electricity consumption statistics of Alaska, Canada, and the United States. The contributions of the Columbia and St Lawrence rivers have, in this table, been divided equally between the United States and Canada. In Alaska where the population, and therefore consumption of electricity, is small, the position is anomalous. In Canada, however, more than half again as much electricity as was consumed in 1979 could be supplied by river power were this resource to be fully developed. The proportion that could be generated by harnessing river power in the United States, where very large quantities of electricity are consumed, is only 11 per cent. For the continent as a whole the possible contribution from river energy is 31 per cent.

River and gauging station		Average Flow $m^3 s^{-1}$	Slope $m\ km^{-1}$	Length km
MACKENZIE	Norman Wells	7340	0.74	4240
OHIO	Metropolis	7279	0.17	2101
ST LAWRENCE	Cornwall	6430	0.12	3379
COLUMBIA	The Dalles	5348	0.75	2000
YUKON	Ruby	4811	0.23	2848
MISSISSIPPI	Alton	2782	0.17	1883
FRAZER	Hope	2710	1.49	1368
MISSOURI	Hermann	2156	0.56	4047
MOBILE	Mobile	1737	0.71	1255
GRANDE RIVIERE	l'Achazi	1730	0.79	650
CHURCHILL	Muskrat Falls	1630	0.33	1690
SAGUENAY	Isle Maligne	1470	1.11	625
ALBANY	Hat Island	1240	0.30	981
ARKANSAS	Little Rock	1192	1.35	2333
OTTOWA	Chats Falls	1170	0.42	1120
NOTTAWAY	Lac Soscumica	1040	0.32	644
SUSQUEHANA	Marietta	992	1.25	714
SKEENA	Usk	916	2.94	579
RED	Alexandria	908	0.58	2043
WINNIPEG	Slave Falls	835	0.22	764
NASS	Schumal Creek	832	5.48	322
WHITE	Clarendon	813	0.42	1158
MOOSE	Moose River	821	0.50	541
CHURCHILL	Granville Falls	797	0.14	1609
SACRAMENTO	Verona	770	2.95	607
ST JOHN	Pokiok	734	0.73	644
ST MAURICE	Grand-Mere	708	0.58	523
SASKATCHEWAN	The Pas	676	1.12	1939
APPALACHICOLA	Chattahoochee	614	1.06	843
COLORADO	Hoover Dam	411	1.33	2252

Figure 6.4 Largest North American Rivers

River	Gross potential PJ	Physical potential PJ
Mackenzie	3570	609
Columbia	1243	770
Frazer	857	197
Missouri	757	96
Arkansas	582	15
Yukon	488	287
St Lawrence	404	112
Ohio	403	131
Skeena	242	23
Mobile	240	23
Nass	228	15
Saskatchewan	227	69
Sacramento	214	124
Colorado	191	125
Red	167	9
Saguenay	158	89
Churchill (Lab.)	141	182
Mississippi	138	5
Grande Riviere	138	296
Susquehana	137	33
Appalachicola	85	13
White	63	13
Ottowa	85	40
Albany	57	26
St John	53	20
North America	10,868	3322

Figure 6.5   Most Energetic North American Rivers

	Electricity consumption 1979 PJ	River energy potential PJ	Percentage
Alaska	10	287	2780
Canada	1242	2007	161
United States	9486	1028	11
North America	10,738	3322	31

Figure 6.6   River Energy in North America

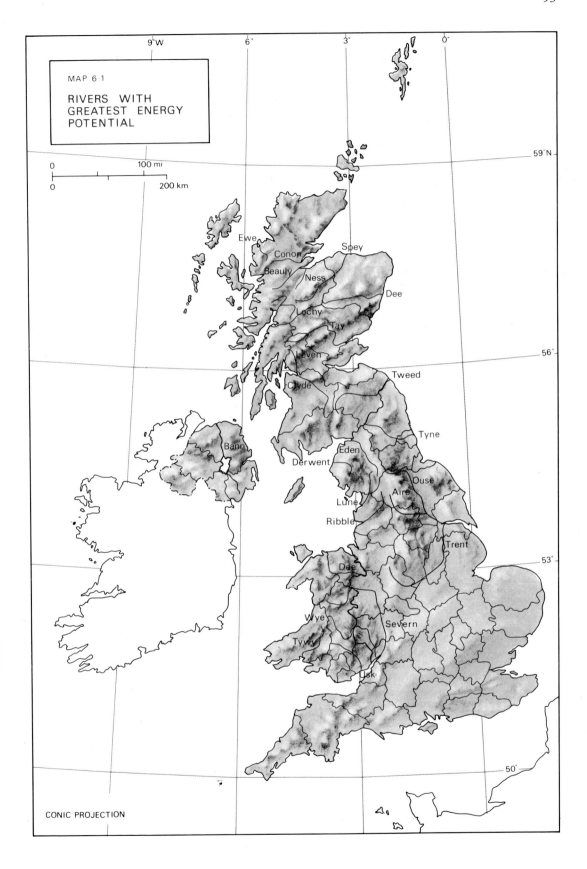

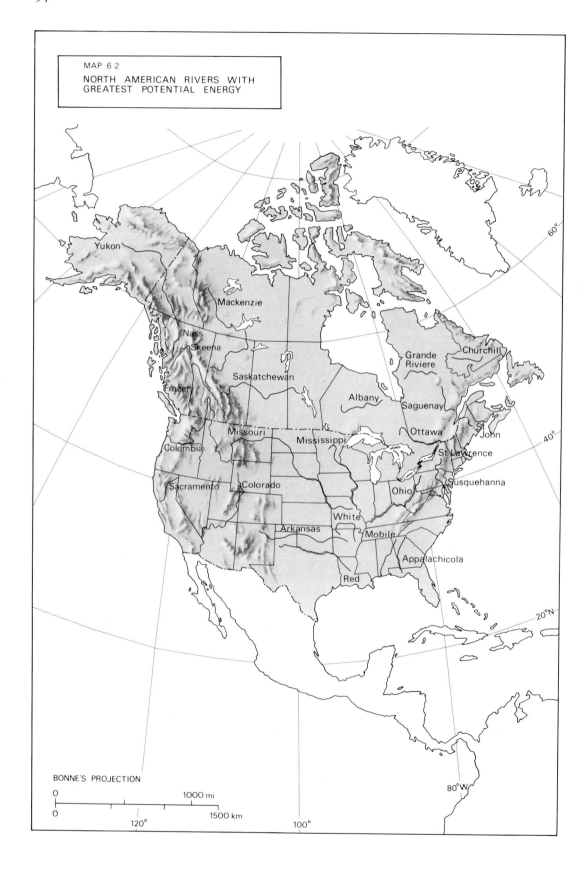

Chapter Seven
Biofuels

THE ORIGIN OF BIOFUELS

More than 99·9 per cent by weight of the biomass, the name given to the Earth's freight of living matter, is composed of green plants. In size plants vary from microscopic algae to the Sequoia Gigantea growing in the forests of the Sequoia National Park in California, where a large specimen of the tree can weigh as much as 2000 tonnes.

Every green plant contains within some of its cells minute sac-like organelles known as chloroplasts. Although so small as to be barely perceptible under an optical microscope, the chloroplasts of a plant possess an intricate internal structure. They also contain the green organic compound chlorophyll whose colour gives this division of the living world its general name. The chloroplasts of green plants are the factories of the substance of life, for within them are manufactured by the process of photosynthesis the organic compounds from which the material of the entire living world is derived.

The process of photosynthesis, by means of which life on Earth is sustained, is not yet understood in full detail. It is known, however, that as a result of a complicated series of interrelated chemical reactions occurring within the chloroplasts, carbohydrates are formed from water and carbon dioxide. The overall equation can be represented as:

$$6H_2O + 6CO_2 + 4\cdot 66 \times 10^{-18} J \to C_6H_{12}O_6 + O_2$$

The energy required for the photosynthetic reaction to operate is furnished by sunlight. The glucose and the free oxygen to the right of the arrow have a higher energy content than is possessed by the carbon dioxide and water forming the left side of the equation. Through photosynthesis, therefore, solar energy is sequestered in the substance of green plants and also, by the food chains which link together the biosphere, in the tissues of all living things. The energy contained in organic materials can be liberated by the converse process of combustion, in which oxygen is consumed and carbon dioxide and heat given out when organic matter is oxidized. Those organic materials which are plentiful and accessible enough to produce useful quantities of energy when burnt or processed are known collectively as biofuels.

The rate at which photosynthesis proceeds, and therefore the rate of production of organic material, will vary with the plant's circumstances. The most influential factors governing the speed of plant growth are the availability of solar energy, water, carbon dioxide, nitrogen, phosphorous, and various other minerals. Influences which restrain photosynthetic activity are low temperatures, poor soil quality, and man-made or naturally occurring pollution. In the conditions prevailing in the United Kingdom plants achieve the average rates of production listed in Figure 7.1. The figures refer to the total of organic material produced, not just that part which is at present economically significant.

MATERIALS FOR BIOFUELS

The surface of the Earth, upon which the plants and animals constituting the biomass reside, can itself be regarded as a natural resource. In a country as densely populated as the United Kingdom,

	Productivity dry t ha^{-1} yr^{-1}
Spring wheat	10
Barley	11
Fodder grass	20
Potatoes	11
Sugar beet	23
Kale	21

Figure 7.1   Plant Productivity

where a land area of 24,093,000 ha supported a human population in 1979 of 55,883,000, land is in short supply and has for centuries been regarded as a precious commodity to be utilized intensively. Almost exactly half the land area of the United Kingdom is devoted to crops or permanent grass. Rough grazing land, which makes up a further 28 per cent of the United Kingdom total, and which is popularly supposed to be in a natural state, would revert to forest or scrub were the teeth of man's sheep and cattle to be removed. Even moorland above the tree line, often thought of as waste, is managed for the support of deer or game birds. Consequently in Britain no land remains in its primaeval state, untouched by human activity, to form a reserve of space for colonization. Land use in the United Kingdom is classified in Figure 7.2.

	ha × 10^6	Percentage
Cropland	7·227	30·0
Permanent grass	4·926	20·4
Rough grazing	6·678	27·7
Woodland	1·908	7·9
Urban land	1·918	8·0
Other	1·436	6·0
Total	24·093	100·0

Figure 7.2   Land Use in the United Kingdom 1972

Despite the fact that cropland, grassland, and woodland together occupy 86 per cent of the area of the United Kingdom, it remains that the British agricultural and forestry industries can supply only a part of the food, fibre, and industrial raw material needed by the nation's economy. More than 90 per cent of the timber consumed and nearly half the country's supply of food is imported from abroad. No doubt future improvements in agricultural productivity will bring about a reduction in the scale of food imports.

Existing British forests are mostly young and can be expected to become more productive as, with the passage of time, they mature. Furthermore, the existing forest estate could be nearly doubled in extent. But, when all likely improvements are allowed for, it remains that a large increase in domestically-produced supplies of food and materials is impossible because there exists no untapped resource of space for the expansion of forestry or farming. For the foreseeable future the nation's economy will require large imports and Britain will continue to be hard pressed against her small resources of land. It is unlikely that more than an insignificant amount of space will ever become available in the United Kingdom for growing energy crops, and energy plantations are therefore not taken into consideration in this chapter.

Not all the material produced by agriculture and forestry reaches the consumer, nor can it all be used by the economy as it is at present structured. Nearly half a tree, for instance, remains as waste material when a forest is felled. Cereal straw and the leaves and stalks of many green crops are disposed of as waste, while accumulations of livestock manure are regarded as a problem in waste disposal and pollution control. These organic materials, at present regarded as wastes, are in fact repositories of photosynthetically-manufactured energy. They can be gathered and used, without increasing the pressure on land resources, to supply considerable amounts of useful energy.

BIOMASS CONVERSION METHODS

Several processes are available for the purpose of tapping the energy content of organic materials and converting them into useful fuels. Oxidation of biofuels by combustion as a method of producing heat is employed throughout the world, often at very low levels of thermal efficiency. At the other pole of sophistication from simple burning is the process of pyrolysis, in which organic feedstock is decomposed in a reactor by heating it in the absence of air. Pyrolysis can take place at temperatures from 500 °C to 700 °C and has the advantage of producing liquid and gaseous fuels of high calorific value. The reactor and its associated equipment is complicated and expensive, however, and the plant does not lend itself to small scale installations. Intermediate in complexity between combustion and pyrolysis are the processes of gasification and anaerobic digestion.

Relatively dry organic matter, such as wood or straw, is a suitable feedstock for gasifying reactors. These installations are constructed to burn part of the feedstock while converting the rest by both pyrolysis and oxidation, into a useful biogas. Oxygen gasification, in which controlled quantities of oxygen are supplied to support combustion, produces a gas containing about 35 per cent hydrogen and methane and which has a calorific value of $12\,\text{MJ}\,\text{m}^{-3}$. Both this type of gas and the more calorific gas produced by anaerobic digestion are grouped in this chapter under the description of biogas.

Chemical methods of conversion such as gasification are not adapted to accept organic materials with a high moisture content. The roots and stalks of green plants, roots and tubers such as sugar beet and potatoes, and animal manures may contain as much as 90 per cent water. Feedstock of this kind can be converted into gas by the biological process of anaerobic digestion. In the correct conditions of temperature and composition a digester tank containing organic material suspended in water will acquire a population of anaerobic bacteria. These feed upon the tank contents and convert part of the feedstock into a gas. Biogas produced by anaerobic digestion is about 50 per cent methane and has a calorific value of $27\,\text{MJ}\,\text{m}^{-3}$. The sludge remaining after the process is complete retains its value as a fertilizer and can usefully be returned to the land.

In this chapter it is assumed that forestry wastes and cereal straw will be converted to gas by oxygen gasification. Anaerobic digestion is assumed to be the process employed to make biogas from the wastes of tillage crops and from animal manures. Since water and other inorganic chemicals are not converted to a fuel when processed by gasification or digestion the rate of energy production of any biomass conversion system is dependent upon the proportion of the feedstock that is composed of organic material. The organic components of the feedstock are referred together as volatile solids and are abbreviated VS in Figures 7.4, 7.6, 7.8, 7.10, and 7.12.

Although natural gas, obtained from gas fields under the North Sea, is now distributed throughout the United Kingdom by a system of underground mains, it is unlikely that biogas could be distributed to the customer in the same nationwide manner. It has been seen that biogas has a calorific value of $12\,\text{MJ}$ or $27\,\text{MJ}\,\text{m}^{-3}$ according to its method of production.

Natural gas, at $35\,\text{MJ}\,\text{m}^{-3}$ has a high enough energy density to be worth distributing over long distances. Biogas, however, would be very expensive in money and energy to move such long distances and for this reason the descriptions of the biofuels resources which follow are arranged on a regional basis. It will be found that regional biofuel potentials, and the local levels of demand for energy in the form of gas, are subject to wide variations. A few regions emerge as particularly favourable areas for the exploitation of biofuels as a renewable energy resource, while elsewhere biofuels are too scarce to make more than a relatively minor contribution to gas supplies.

FORESTRY WASTES

In 1979 productive woodland occupied 1,686,700 ha and covered 6·9 per cent of the land area of the United Kingdom. In Europe only the Irish Republic is more treeless than the United Kingdom. Not only is the nation's tree cover small in extent but in the years between 1939 and 1945 the requirements of the war effort brought about the premature felling of much of the existing stock of growing timber. Consequently the forest estate of 1979 is young and has yet to reach its full productive potential. In 1977 each hectare of British woodland produced only $1\cdot6\,\text{m}^3$ of timber while the larger and older forests of France, Germany, and Belgium produced $2\cdot2$, $4\cdot0$, and $4\cdot4\,\text{m}^3$ respectively. Figure 7.3 describes by regions the size of the existing and the possible United Kingdom forest estate. Also shown in Figure 7.3 is the area of forest that would be possessed by the nation if, over the next 50 years, a vigorous tree planting programme were pursued and brought to fruition. All land marginal to agriculture and not needed for other purposes, but suitable for timber growing, would upon completion of the programme be afforested. The forest estate would then be more than doubled to a total area of 3,646,700 ha, placing 15 per cent of the country under forest cover. By the year 2030 British forests would be approaching

	Area ha	Biogas PJ		Area ha	Biogas PJ
Scotland present 1979 maximum 2030	713,700 2,034,100	2·99 26·87	*East Anglia* present 1970 maximum 2030	60,400 75,400	0·25 0·99
North present 1979 maximum 2030	129,900 277,200	0·54 3·66	*Southwest* present 1979 maximum 2030	140,900 286,000	0·59 3·78
Yorks & H'side present 1979 maximum 2030	47,600 185,000	0·19 2·44	*Southeast* present 1979 maximum 2030	223,000 381,100	0·94 5·03
Northwest present 1979 maximum 2030	22,700 88,000	0·09 1·16	*Wales* present 1979 maximum 2030	194,200 513,700	0·81 6·78
W Midlands present 1979 maximum 2030	49,900 130,100	0·21 1·72	*N Ireland* present 1979 maximum 2030	65,000 185,100	0·27 2·45
E Midlands present 1979 maximum 2030	39,500 93,800	0·16 1·24	*United Kingdom* present 1979 maximum 2030	1,686,700 3,646,700	7·08 48·17

Figure 7.3 Productive Woodland and Biogas Potential

maturity and an average productivity of 5 m³ ha⁻¹ yr⁻¹ could be expected from them. This scale of afforestation and timber production can be achieved at no cost to the nation's economy and without making inroads into land for which there are more than marginally productive alternative uses.

Both the present area of productive woodland and the size of the potential United Kingdom forest estate are shown on Map 7.1. It will be seen that the greatest quantity of timber, and also the largest scope for forest expansion, exists in Scotland. Northern England, Wales, southwest England, and southeast England also possess considerable stands of timber while Northern Ireland, the Midlands, Yorkshire and Humberside, and East Anglia are the most treeless parts of the country. East Anglia, where nearly all available land is farmed, has the smallest scope for the expansion of forestry.

The symbols on the maps accompanying this chapter, including those illustrating the areas of the country under timber, have been drawn at the scale of the map on which they appear. The shape of the symbol does not, of course, respresent the configuration of the type of land use in question but symbols are correctly sized relative to one another and to their map.

When a forest is felled for timber some 45 per cent of the tree remains behind in the form of leaves, branches, and roots. If slightly less than half these wastes were to be gathered and processed by oxygen gasification, the gas productivity and calorific output shown in the last two columns of Figure 7.4 could be achieved. It will be noted that the mature forests to be foreseen in the second quarter of the next century would be more than three times as productive of energy as are existing British woodlands.

Totals of the energy available from forestry wastes are obtained by multiplying forest areas by the appropriate factor from Figure 7.4. The results appear in the second column of Figure 7.3, where it will be seen that although existing woodlands would yield only just over 7 PJ of energy each year an annual total of nearly 50 PJ of biogas could be expected in the next century from an enlarged United Kingdom forest estate.

	Harvest		Recoverable wastes		Gas productivity m³ t⁻¹ VS	Gas production m³ ha⁻¹	Calorific value GJ ha⁻¹ (12 MJ m⁻³)
	m³ ha⁻¹	wet tonnes ha⁻¹ (1 m³ =2·2 t)	wet tonnes ha⁻¹ (20% of harvest)	VS t ha⁻¹ (70% wet wt)			
Present estate	1·60	3·52	0·70	0·49	715·00	350·35	4·20
Maximum estate	5·00	11·00	2·20	1·54	715·00	1101·10	13·21
U.S.A.	1·70	3·74	0·75	0·53	715·00	378·95	4·55
Canada	0·38	0·84	0·17	0·12	715·00	85·80	1·03

Figure 7.4 Oxygen Gasification of Forestry Wastes

CEREAL STRAW

Cereal production occupied 3,759,400 ha in the United Kingdom in 1979, accounting for 15·4 per cent of the land area of the country. This is nearly twice the area devoted to all woodland. The three cereal crops grown in Britain are, in ascending order of importance, oats, wheat, and barley. Barley is grown in all parts of the country although, as shown on Map 7.2, areas under barley in Wales, northwest England, and Northern Ireland are smaller than elsewhere. Wheat is a cereal grown mainly in lowland Britain with only small quantities sown in northern and western regions. Oats, the distribution of whose cultivation is shown in Map 7.4, is a considerable crop in Scotland but of minor importance elsewhere. Areas under cereals in the 11 regions of the United Kingdom are listed in Figure 7.5.

When at harvest the ripe grain is collected for food a residue remains in the fields as straw and chaff. The quantity of straw involved depends upon the type of cereal and the height above the ground at which it is cut during reaping. Average straw arisings in Britain are given in the first

	Area ha	Biogas PJ		Area ha	Biogas PJ
Scotland			*East Anglia*		
winter wheat	20,900	0.41	winter wheat	267,900	5.24
winter barley	425,600	6.86	spring wheat	8,600	0.11
oats	46,100	0.42	winter barley	87,900	1.42
		7.69	spring barley	157,600	1.75
North			oats	5,800	0.05
winter wheat	24,600	0.48			8.57
spring wheat	800	0.01	*Southwest*		
winter barley	11,300	0.18	winter wheat	115,800	2.26
spring barley	112,200	1.25	spring wheat	3,600	0.05
oats	6,800	0.06	winter barley	90,300	1.45
		1.98	spring barley	159,300	1.77
Yorks & H'side			oats	14,600	0.13
winter wheat	130,400	2.55			5.66
spring wheat	4,000	0.05	*Southeast*		
winter barley	38,500	0.62	winter wheat	363,700	7.11
spring barley	200,700	2.23	spring wheat	11,200	0.14
oats	10,700	0.10	winter barley	152,800	2.46
		5.55	spring barley	221,700	2.46
Northwest			oats	23,800	0.22
winter wheat	8,000	0.16			12.93
spring wheat	200	0.01	*Wales*		
winter barley	8,100	0.13	winter wheat	5,000	0.09
spring barley	49,300	0.55	spring wheat	200	0.00
oats	1,600	0.01	winter barley	18,100	0.29
		0.86	spring barley	40,400	0.45
W Midlands			oats	11,100	0.10
winter wheat	101,700	1.99			0.93
spring wheat	3,100	0.04	*N Ireland*		
winter barley	54,500	0.88	winter wheat	500	0.01
spring barley	111,000	1.23	spring wheat	—	0.00
oats	12,500	0.11	winter barley	7,800	0.12
		4.25	spring barley	44,500	0.49
E Midlands			oats	4,200	0.04
winter wheat	267,400	5.23			0.66
spring wheat	8,500	0.11	*United Kingdom*		
winter barley	89,600	1.44	winter wheat	1,305,900	25.54
spring barley	182,200	2.03	spring wheat	40,200	0.52
oats	12,700	0.12	winter barley	984,500	15.86
		8.93	spring barley	1,278,900	14.22
			oats	149,900	1.38
				3,759,400	57.52

Figure 7.5 Cereal Areas and Biogas Potential

column of Figure 7.6 from which it will be seen that a hectare of winter wheat produces more than twice the weight of straw that is available from the same area under oats. Some straw is at present put to use as livestock bedding but by far the larger part of the straw crop is disposed of as a waste product by burning. About 12 million tons of straw are available from cereal growing in the United Kingdom each year and it is of interest to investigate the energy potential of this renewable biofuel resource.

Figure 7.6 give the energy that could be obtained from six cereal crops if the recoverable part of their straw arisings were to be converted to biogas by oxygen gasification. The calorific value of the biogas obtainable from straw is found by multiplying the area of each cereal crop by the appropriate factor from Figure 7.6. Figure 7.5 shows that the calorific value of United Kingdom straw arisings in 1979 is greater than that of forestry wastes even if our forests were to be doubled in extent during the course of the next 50 years.

	Arisings	Recoverable wastes		Gas productivity $m^3\ t^{-1}$ VS	Gas production $m^3\ ha^{-1}$	Calorific value $GJ\ ha^{-1}$ ($12\ MJ\ m^{-3}$)
	Total air dried tonnes ha^{-1}	air dried tonnes ha^{-1} (75% of arisings)	VS tonnes ha^{-1} (80% air dried wt)			
Winter wheat	3·80	2·85	2·28	715·00	1630·20	19·56
Spring wheat	2·49	1·87	1·50	715·00	1072·50	12·87
Winter barley	3·13	2·35	1·88	715·00	1344·20	16·13
Spring barley	2·16	1·62	1·30	715·00	929·50	11·15
Oats	1·79	1·34	1·07	715·00	765·05	9·18
Maize	6·50	4·88	3·90	715·00	2788·50	33·46

Figure 7.6   Oxygen Gasification of Cereal Straw

TILLAGE CROPS

A wide variety of vegetables are grown in Britain but only four are present in sufficient quantity and produce enough organic residue after harvest to be of interest as biofuel feedstock. The areas devoted to sugar beet, brussel sprouts, peas, and potatoes are listed by region in Figure 7.7 and shown on Map 7.5. It will be seen that these crops are mainly grown in the eastern part of the country. Arisings from tillage crops are given in Figure 7.8.

Waste materials from tillage crops have a high water content and are consequently unsuitable for conversion to biogas by oxygen gasification. The process of anaerobic digestion is, however, well suited to wet feedstocks of this type. The calorific value of tillage crops can be obtained by multiplying the areas devoted to each crop by the appropriate factor in Figure 7.8. The results are given in the second column of Figure 7.7. It will be noted that sugar beet wastes are by far the most important tillage crop biofuel resource.

CATCH CROPS

In the United Kingdom the harvest of winter wheat and winter barley is in an average year completed by the beginning of August. Spring sown cereals mature later and their harvest is not usually in until the end of August. Planting of the next crop of winter cereal is generally carried out in November with the result that the land lies bare for as long as three months in late summer and autumn. This later part of the growing season can be exploited by sowing, at the time of cereal harvest, a second or catch crop grown for fodder or as a biofuel feedstock. Catch crops enable the existing area of cultivated land to be utilized more intensively.

	Area ha	Biogas PJ		Area ha	Biogas PJ
Scotland			*East Anglia*		
sugar beet	nil	0·00	sugar beet	105,490	5·65
brussel sprouts	400	0·01	brussel sprouts	750	0·02
peas	3,320	0·06	peas	15,140	0·27
potatoes	33,080	0·27	potatoes	26,250	0·21
		0·34			6·15
North			*Southwest*		
sugar beet	nil	0·00	sugar beet	590	0·03
brussel sprouts	nil	0·00	brussel sprouts	nil	0·00
peas	60	0·00	peas	100	0·00
potatoes	4,260	0·03	potatoes	7,380	0·06
		0·03			0·09
Yorks & H'side			*Southeast*		
sugar beet	29,460	1·58	sugar beet	7,680	0·41
brussel sprouts	660	0·02	brussel sprouts	250	0·01
peas	5,490	0·10	peas	4,410	0·08
potatoes	23,920	0·19	potatoes	16,200	0·13
		1·89			0·63
Northwest			*Wales*		
sugar beet	980	0·05	sugar beet	nil	0·00
brussel sprouts	30	0·00	brussel sprouts	50	0·00
peas	10	0·00	peas	30	0·00
potatoes	8,080	0·07	potatoes	3,780	0·03
		0·12			0·03
W Midlands			*N Ireland*		
sugar beet	19,230	1·03	sugar beet	nil	0·00
brussel sprouts	50	0·00	brussel sprouts	100	0·00
peas	100	0·00	peas	nil	0·00
potatoes	11,440	0·09	potatoes	12,590	0·10
		1·12			0·01
E Midlands			*United Kingdom*		
sugar beet	50,120	2·68	sugar beet	213,550	11·44
brussel sprouts	1,560	0·05	brussel sprouts	3,850	0·12
peas	17,600	0·31	peas	46,260	0·83
potatoes	29,290	0·24	potatoes	176,270	1·44
		3·28			13·83

Figure 7.7 Tillage Crop Areas and Biogas Potential

	Arisings	Recoverable wastes		Gas productivity $m^3\ t^{-1}$ VS	Gas production $m^3\ ha^{-1}$	Calorific value GJ ha^{-1} (12 MJ m^{-3})
	Total wet tonnes ha^{-1}	wet tonnes ha^{-1} (66% of arisings)	VS tonnes ha^{-1} (20% wet wt)			
Sugar beet, leaf and crown	30·00	19·80	3·96	501	1983·96	53·57
Brussel sprouts, stalks	16·00	10·56	2·11	557	1175·27	31·73
Peas, haulms	9·00	5·94	1·19	557	662·83	17·90
Potatoes, chats	3·75	2·48	0·50	606	303·00	8·18

Figure 7.8 Anaerobic Digestion of Tillage Crop Wastes

Figure 7.9 list the areas of land by region that would be available for catch crops in a year such as 1979. In addition to the area under cereal the totals include land used for growing rape since that crop is harvested at the same time of the year as are cereals. Map 7.6 shows that catch crops could be grown in all parts of the country but that the contribution they could make in Wales, northwest England, and in Northern Ireland is relatively small.

The calorific value of the biogas to be expected from catch crops is obtained by multiplying the available land area by the value in the last column of Figure 7.10. The results are shown in the second column of Figure 7.9. It will be noted that, although catch crops are a smaller biofuel resource than cereal straw, their contribution could be more than three times as great as the energy obtainable from forestry wastes in 1979.

	Area ha	Biogas PJ		Area ha	Biogas PJ
Scotland			*East Anglia*		
cereals	492,600	2·93	cereals	527,800	3·14
rape	7,700	0·04	rape	13,600	0·08
		2·97			3·22
North			*Southwest*		
cereals	155,700	0·92	cereals	383,600	2·28
rape	2,000	0·01	rape	4,800	0·03
		0·93			2·31
Yorks & H'side			*Southeast*		
cereals	384,300	2·29	cereals	773,200	4·60
rape	7,700	0·04	rape	23,400	0·13
		2·33			4·73
Northwest			*Wales*		
cereals	67,200	0·39	cereals	74,800	0·44
rape	2,300	0·01	rape	6,200	0·04
		0·40			0·48
W Midlands			*N Ireland*		
cereals	282,800	1·68	cereals	57,000	0·34
rape	3,800	0·02	rape	nil	0·00
		1·70			0·34
E Midlands			*United Kingdom*		
cereals	560,400	3·33	cereals	3,759,400	22·37
rape	26,000	0·15	rape	97,500	0·58
		3·48		3,856,900	22·95

Figure 7.9 Catch Crop Areas and Biogas Potential

	Arisings	Recoverable material		Gas productivity $m^3\ t^{-1}$ VS	Gas production $m^3\ ha^{-1}$	Calorific value $GJ\ ha^{-1}$ (12 MJ m^{-3})
	Total wet tonnes ha^{-1}	wet tonnes ha^{-1} (66% of arisings)	VS tonnes ha^{-1} (20% wet wt)			
Kale, whole plant	3·00	1·98	0·40	557	220·57	5·95

Figure 7.10 Anaerobic Digestion of Catch Crops

LIVESTOCK MANURE

In 1979 permanent grassland occupied 4,926,000 ha or 20·4 per cent of the land area of the United Kingdom. A further 6,678,000 ha is classified as rough grazing. The fodder grass grown on this land, together with that furnished by cropland under grass, provides much of the food needed by the nation's population of more than 13 million cattle, nearly 8 million pigs, and over 112 million chickens. Part of the manure produced by livestock is by present practice returned to the land as fertilizer but most is disposed of as an agricultural waste product. In intensive rearing units, particularly, accumulations of manure are regarded as a nuisance and a source of pollution. But the organic content of manure is high and it is a suitable feedstock for the production of biogas by anaerobic digestion. The following paragraphs assess livestock manures as a biofuel resource.

Figure 7.11 lists the United Kingdom cattle, pig, and poultry populations by region. Sheep are excluded from this assessment because they are never housed and their manure is too dispersed to be collected for use as a biofuel feedstock. Cattle numbers are highest in the western parts of the country as shown on Map 7.7. Scotland, Northern Ireland, Wales, and southwest England support 57 per cent of the nation's cattle while the largely arable region of East Anglia has only 2 per cent of the total. Pigs, on the other hand, are more numerous in the eastern regions while poultry numbers are fairly uniformly distributed across the country. Maps 7.9 and 7.10 show that relatively few laying hens are kept in Wales and northern England and that north and northwest England breed fewer broilers than other regions.

Manure arisings, gas productivity, and calorific equivalents per animal are tabulated in Figure 7.12. It is supposed, in the second column of this table, that only one-third of beef cattle manure is recoverable because beef herds spend much time at pasture. Other livestock is housed for all or most of the time so enabling the manure to be collected and used.

The biogas totals by region given in the second column of Figure 7.11 are obtained by multiplying animal populations by the appropriate factor from Figure 7.12. It will be noted that the energy obtainable from animal manure is greater than that which could be supplied by forestry wastes, cereal straw, tillage crops, or catch crops. The most important sources of manure for biofuels are dairy cows, capable of supplying 37·8 PJ annually, and young cattle, from whose wastes 23·27 PJ could be obtained each year.

THE SIZE OF THE RESOURCE

Calorific values of the biogas available from the five resources discussed in this chapter are tabulated in Figure 7.13. Two totals of the quantity of energy to be expected are given. The first column includes only the gas production potential of the small area under forest in 1979, while the higher total in the second column takes account of the much larger amount of energy obtainable from the 3·6 million ha forest estate possible half a century later. The national totals of 185·82 PJ and 226·91 PJ in 1979 and 2030 conceal large variations in the energy potentials of the regions of the United Kingdom and between the relative importance of the different biofuel resources within those regions.

Forestry wastes in 1979 could supply as much as 12 per cent of biogas production only in Scotland. Wales, where forestry could provide 8 per cent, and northern England, where 6 per cent of the biogas total could come from this resource, are the only other areas of the country where the present forest estate could supply more than relatively small amounts of energy. In the year 2030, however, the energy significance of forestry could be much greater. By then Scotland could obtain 54 per cent of its biogas from forestry while Wales, northern England, Northern Ireland, southeast England, and southwest England could expect proportions of 42 per

	Population	Biogas PJ		Population	Biogas PJ
Scotland			*East Anglia*		
cattle 0–2 yrs	1,408,400	4·84	cattle 0–2 yrs	169,800	0·58
beef cattle	659,200	2·05	beef cattle	51,700	0·16
dairy cows	373,900	3·56	dairy cows	90,100	0·86
pigs	521,500	0·46	pigs	1,368,700	1·21
laying hens	4,317,700	0·44	laying hens	3,516,600	0·36
broilers	5,861,300	0·26	broilers	12,219,700	0·55
		11·61			3·72
North			*Southwest*		
cattle 0–2 yrs	406,400	1·39	cattle 0–2 yrs	1,065,300	3·66
beef cattle	229,100	0·71	beef cattle	337,600	1·05
dairy cows	257,700	2·46	dairy cows	944,000	9·01
pigs	162,700	0·14	pigs	857,800	0·76
laying hens	1,344,500	0·14	laying hens	7,169,700	0·73
broilers	2,334,900	0·11	broilers	6,417,300	0·29
		4·95			15·50
Yorks & H'side			*Southeast*		
cattle 0–2 yrs	416,500	1·43	cattle 0–2 yrs	595,100	2·04
beef cattle	131,400	0·41	beef cattle	188,100	0·58
dairy cows	237,500	2·27	dairy cows	388,300	3·71
pigs	1,347,100	1·19	pigs	1,180,900	1·04
laying hens	3,048,700	0·31	laying hens	8,624,800	0·88
broilers	6,214,600	0·28	broilers	8,722,600	0·39
		5·89			8·64
Northwest			*Wales*		
cattle 0–2 yrs	278,300	0·95	cattle 0–2 yrs	694,200	2·38
beef cattle	74,200	0·23	beef cattle	308,800	0·96
dairy cows	340,900	3·25	dairy cows	413,900	3·95
pigs	431,200	0·38	pigs	133,700	0·12
laying hens	3,769,000	0·38	laying hens	1,970,700	0·20
broilers	2,432,300	0·11	broilers	4,102,600	0·18
		5·30			7·79
W Midlands			*N Ireland*		
cattle 0–2 yrs	548,500	1·88	cattle 0–2 yrs	786,500	2·70
beef cattle	163,200	0·51	beef cattle	457,800	1·42
dairy cows	386,400	3·69	dairy cows	296,300	2·83
pigs	479,300	0·42	pigs	721,800	0·64
laying hens	3,427,000	0·35	laying hens	5,171,200	0·53
broilers	4,405,200	0·20	broilers	4,490,400	0·20
		7·05			8·32
E Midlands			*United Kingdom*		
cattle 0–2 yrs	406,200	1·39	cattle 0–2 yrs	6,775,200	23·27
beef cattle	173,300	0·54	beef cattle	2,774,400	8·63
dairy cows	232,500	2·22	dairy cows	3,961,500	37·80
pigs	657,300	0·58	pigs	7,862,000	6·96
laying hens	5,215,900	0·53	laying hens	47,575,800	4·85
broilers	7,987,300	0·36	broilers	65,188,200	2·93
		5·62			84·44

Figure 7.11   Livestock Populations and Biogas Potential

	Arisings wet t	Recoverable wastes				Gas productivity m³ t⁻¹	Gas production m³	Calorific value GJ/animal (27 MJ m⁻³)
		recoverable %	amount recoverable tonnes	VS %	VS t			
Cattle, 0 to 2 years	5·66	75	4·24	10	0·424	300	127·200	3·434
Beef animals, over 2 years	10·22	33	3·37	10	0·337	342	115·254	3·112
Dairy cows, over 2 years	14·96	75	11·22	10	1·122	315	353·430	9·542
Pigs, whole population	0·73	90	0·66	12	0·079	415	32·785	0·885
Laying hens	0·051	90	0·046	20	0·009	416	3·744	0·102
Table fowls	0·023	90	0·021	20	0·004	416	1·664	0·045

Figure 7.12   Anaerobic Digestion of Livestock Manure

cent, 32 per cent, 21 per cent, 16 per cent, and 14 per cent respectively. These considerations show that forestry is an important future energy resource, particularly for the northern, western, and southern regions of the country.

Cereal straw was in 1979 a larger potential source of biogas than forestry wastes by a factor of more than eight. The four eastern regions of the country, Yorkshire and Humberside, East Midlands, East Anglia, and the Southeast, account for 63 per cent of the gas obtainable from cereal straw. A similar distribution of production is to be seen in the potential of tillage and catch crops where the same four regions account for 86 per cent and 60 per cent of the national totals. Cereals, tillage crops, and catch crops together account for just over half the energy that could be obtained from biofuels in 1979, and their relative importance in the south and east reflects the predominance of arable farming in those parts of the country.

Although Scotland, Northern Ireland, Wales, and southeast England could produce 51 per cent of the biogas obtainable from livestock manure, the regional distribution of this resource is more even than is the case with forestry wastes, cereal straw or tillage, and catch crop residues. Map 7.7 shows that the cattle population is high in northern and western regions where soil conditions and climate favour grassland farming. Numbers of pigs and fowls, however, are higher in the east and south. This is the result of modern methods of intensive rearing in which feed is brought to the animal rather than stock being put out to forage or graze.

The usefulness, as opposed to the simple availability, of the biofuels resource is illustrated on Map 7.11 and tabulated in Figure 7.14. In 1979 1723·6 PJ of energy were sold to consumers in the United Kingdom in the form of mains gas. This level of demand may be used as a means of measuring the size of the biofuels energy resource. The 1979 biogas potential of 185·82 PJ is only 11 per cent of the sales of mains gas in that year, while only 13 per cent would be available if the potential of a larger forest estate in the year 2030 is taken into account. It must therefore be concluded that, from a national point of view, biofuels could at present make only a marginally significant contribution to the nation's supply of gas. Nevertheless, they have an important future role to play in filling an energy gap which is expected to appear in Britain when existing reservoirs of natural gas are exhausted. The place of biogas in the economy of an energy-conscious future is discussed in the concluding chapter of this atlas and is illustrated in Figure 10.2.

In some parts of the country the biofuels potential is large in relation to the local level of demand. Map 7.11 and Figure 7.14 show that in Scotland the 1979 and 2030 biogas potentials as a proportion of 1979 mains gas consumption could amount to 27 per cent and 52 per cent respectively. Other areas possessing useful biogas resources are East Anglia and southwest England. The very low level of mains gas consumption in Northern Ireland produces an anomalous result in the province. But even were gas consumption in Ulster to as much as quintuple when natural gas becomes available through the mains system, the biogas potential of the region would still stand at 60 per cent rising to 74 per cent by the year 2030.

	Present forest PJ	Maximum forest PJ		Present forest PJ	Maximum forest PJ
Scotland			*East Anglia*		
forestry wastes	2·99	26·87	forestry wastes	0·25	0·99
cereal straw	7·69	7·69	cereal straw	8·57	8·57
tillage crops	0·34	0·34	tillage crops	6·15	6·15
catch crops	2·97	2·97	catch crops	3·22	3·22
livestock manure	11·61	11·61	livestock manure	3·72	3·72
	25·60	49·48		21·91	22·65
North			*Southwest*		
forestry wastes	0·54	3·66	forestry wastes	0·59	3·78
cereal straw	1·98	1·98	cereal straw	5·66	5·66
tillage crops	0·03	0·03	tillage crops	0·09	0·09
catch crops	0·93	0·93	catch crops	2·31	2·31
livestock manure	4·95	4·95	livestock manure	15·50	15·50
	8·43	11·55		24·15	27·34
Yorks & H'side			*Southeast*		
forestry wastes	0·19	2·44	forestry wastes	0·94	5·03
cereal straw	5·55	5·55	cereal straw	12·93	12·93
tillage crops	1·89	1·89	tillage crops	0·63	0·63
catch crops	2·33	2·33	catch crops	4·73	4·73
livestock manure	5·89	5·89	livestock manure	8·64	8·64
	15·85	18·10		27·87	31·96
Northwest			*Wales*		
forestry wastes	0·09	1·16	forestry wastes	0·81	6·78
cereal straw	0·86	0·86	cereal straw	0·93	0·93
tillage crops	0·12	0·12	tillage crops	0·03	0·03
catch crops	0·40	0·40	catch crops	0·48	0·48
livestock manure	5·30	5·30	livestock manure	7·79	7·79
	6·77	7·84		10·04	16·01
W Midlands			*N Ireland*		
forestry wastes	0·21	1·72	forestry wastes	0·27	2·45
cereal straw	4·25	4·25	cereal straw	0·66	0·66
tillage crops	1·12	1·12	tillage crops	0·01	0·01
catch crops	1·70	1·70	catch crops	0·34	0·34
livestock manure	7·05	7·05	livestock manure	8·32	8·32
	14·33	15·84		9·60	11·78
E Midlands			*United Kingdom*		
forestry wastes	0·16	1·24	forestry wastes	7·08	48·17
cereal straw	8·93	8·93	cereal straw	57·52	57·52
tillage crops	3·28	3·28	tillage crops	13·83	13·83
catch crops	3·48	3·48	catch crops	22·95	22·95
livestock manure	5·62	5·62	livestock manure	84·44	84·44
	21·47	22·55		185·82	226·91

Figure 7.13 Size of Biofuel Resources

Those parts of the country into which the human population is concentrated use very large amounts of mains gas. The North, Northwest, Yorkshire and Humberside, West Midlands, and Southeast account for 74 per cent of mains gas sold in 1979. In these regions biogas could provide only a small proportion of the present demand for energy in the form of gas.

	Potential biogas PJ	Mains gas 1979 PJ	Biogas as % of mains gas
Scotland			
present forest estate	25·60	94·50	27·1
maximum forest estate	49·48		52·4
North			
present forest estate	8·43	221·60	3·8
maximum forest estate	11·55		5·2
Yorks & H'side			
present forest estate	15·85	177·10	8·9
maximum forest estate	18·10		10·2
Northwest			
present forest estate	6·77	225·30	3·0
maximum forest estate	7·84		3·5
W Midlands			
present forest estate	14·33	176·30	8·1
maximum forest estate	15·84		9·0
E Midlands			
present forest estate	21·47	128·90	16·7
maximum forest estate	22·55		17·5
East Anglia			
present forest estate	21·91	38·60	56·8
maximum forest estate	22·65		58·7
Southwest			
present forest estate	24·15	95·30	25·3
maximum forest estate	27·34		28·7
Southeast			
present forest estate	27·87	482·10	5·8
maximum forest estate	31·96		6·6
Wales			
present forest estate	10·04	80·70	12·4
maximum forest estate	16·01		19·8
N Ireland			
present forest estate	9·60	3·20	300·0
maximum forest estate	11·78		368·1
United Kingdom			
present forest estate	185·82	1723·60	10·8
maximum forest estate	226·91		13·2

Figure 7.14 Biofuels Totals in the United Kingdom

BIOFUELS IN NORTH AMERICA

The North American continent is very much less densely populated than is the United Kingdom. Every human inhabitant of the United Kingdom has an average of 0·43 ha of land at his disposal while in North America the corresponding figure is 7·59 ha. The population of the United Kingdom may therefore be said to be nearly 18 times as crowded as are the inhabitants of Canada and the United States. The disparity is somewhat less if the size of the population is measured against productive, rather than all, land. Large parts of the North American continent consist of desert, swamp, moorland, high mountains, and arctic tundra. Land of this type, classified as other in Figures 7.2 and 7.15, amounts to nearly 47 per cent of North America but only 6 per cent of the surface of the United Kingdom. Areas of productive land per inhabitant in North America and the United Kingdom are 4·0 and 0·41 ha, differing by a factor of 9·8. It follows from the relative spaciousness of North America that the quantity of biofuels feedstock available for conversion into useful gas is much greater in proportion to the size of the population than it is in the United Kingdom.

	ha × 10^6	Percentage
Cropland, used and idle	214	11·7
Permanent grass	34	1·9
Rough grazing*	175	9·5
Woodland	535	29·1
Urban land	19	1·0
Other	859	46·8
Total	1836	100·0

*includes grazed forest land

Figure 7.15   Land Use in North America

Although nearly 30 per cent of North America is forest, only about 91 per cent of the area under trees is economically productive. The location of productive woodland is shown diagrammatically by region on Map 7.12, from which it can be seen that the Canadian forest estate is about half again as big as that of the United States. The reputation of the Rocky Mountains as a source of timber is illustrated and confirmed by this map, which also shows the great extent of the forests of eastern Canada and the southeastern United States. Perhaps because of the size of the American and Canadian forest estates, and the consequent abundant supplies of timber, the productivity of North American woodlands is much lower than is the case in Europe. The lower rates of timber extraction per hectare are shown in Figure 7.4, where allowance for this factor is made in assessing the biogas potential of the relatively underexploited woodlands of North America.

It has been shown that the two largest cereal crops in the United Kingdom are barley and wheat. In North America, however, barley is a relatively unimportant crop with the leading places being taken by wheat and maize. The regions in which these two crops are grown are shown on Maps 7.13 and 7.14. It will be noted that the Corn Belt justifies its name by its predominant role in the growing of maize, while the three plains regions further west include the greater part of the wheat lands of North America. The distribution of the diary cow population is illustrated on Map 7.15. The regions of Eastern Canada, the Lake States, the Corn Belt, and the Northeast support the largest herds. The Pacific, Appalachian, and Southeast regions are also in a position to utilize large quantities of dairy manure as a biofuel feedstock.

The forest areas, crop areas, and dairy cow populations of each North American region have been multiplied by the appropriate factor from Figures 7.4, 7.6, and 7.12 to produce the statistics set out in Figure 7.16. As in the United Kingdom the importance of the different types of biofuel feedstock varies greatly between regions. Forestry wastes are the only significant biofuel resource in British Columbia, but in Plains North the scarcity of growing timber reduces this resource to a small part of that obtainable from cereal crops. In the Corn Belt maize stover is much the most

	Area ha×10³ or population ×10³	Biogas PJ		Area ha×10³ or population ×10³	Biogas PJ
Alaska			*Plains South*		
forestry wastes	8,782	40·0	forestry wastes	7,163	32·6
maize stover	nil	—	maize stover	540	18·1
wheat straw	nil	—	wheat straw	4,168	51·2
dairy manure	1·3	—	dairy manure	422	4·0
		40·0			105·9
North			*Lake States*		
forestry wastes	25,600	26·4	forestry wastes	12,708	57·8
maize stover	nil	—	maize stover	3,772	126·2
wheat straw	nil	—	wheat straw	1,065	13·1
dairy manure	nil	—	dairy manure	2,660	25·4
		26·4			222·5
British Columbia			*Corn Belt*		
forestry wastes	52,100	53·7	forestry wastes	20,114	91·5
maize stover	11	0·4	maize stover	15,804	528·8
wheat straw	32	0·4	wheat straw	2,392	29·4
dairy manure	110	1·0	dairy manure	1,872	17·9
		55·5			667·6
Canadian Plains			*Delta*		
forestry wastes	73,800	76·0	forestry wastes	13,557	61·7
maize stover	30	1·0	maize stover	31	1·0
wheat straw	10,199	125·3	wheat straw	180	2·2
dairy manure	373	3·6	dairy manure	217	2·1
		205·9			67·0
Eastern Canada			*New England*		
forestry wastes	118,400	122·0	forestry wastes	13,013	59·3
maize stover	434	14·5	maize stover	nil	—
wheat straw	260	3·2	wheat straw	nil	—
dairy manure	1,750	16·7	dairy manure	376	3·6
		156·4			62·9

Figure 7.16   Biofuels Totals in North America

plentiful feedstock whereas in Canada, where the climate is not conducive to maize cultivation, wheat straw constitutes the largest cereal biofuel resource. In no region of the North American continent is dairy manure likely to be more than a relatively minor biofuel feedstock.

The abundance of agricultural wastes in the United States and Canada is reflected in the very large amounts of biogas that a biofuels industry could produce in North America. The biogas total of 2806 PJ a year is nearly two-thirds greater than the 1979 consumption of mains gas by the entire United Kingdom economy. However, the relative importance of the biofuels resource is almost the same in North America as in the United Kingdom. In each case it amounts to 13 per cent of mains gas sales. This follows from the very high levels of gas consumption in certain parts of the United States, notably Texas, California, Louisiana, Illinois, and Ohio. Figure 7.17, which compares biogas potential with mains gas consumption, is illustrated on Map 7.16.

The small amount of mains gas used in the Canadian Maritime Provinces produces an anomalous result there comparable to the biogas situation in Northern Ireland. Elsewhere in North America the potential biogas contribution to supplies varies from 2·3 per cent in Plains South and 3·4 per cent in the Delta region to more than half in Mountain North. The proportionately

	Area ha×10³ or population ×10³	Biogas PJ		Area ha×10³ or population ×10³	Biogas PJ
Maritime Provinces			*North East*		
forestry wastes	11,000	11·3	forestry wastes	15,257	69·4
maize stover	11	0·4	maize stover	1,125	37·6
wheat straw	10	0·1	wheat straw	176	2·2
dairy manure	119	1·1	dairy manure	1,796	17·1
		12·9			126·3
Pacific			*Appalachians*		
forestry wastes	24,646	112·1	forestry wastes	21,287	96·9
maize stover	149	5·0	maize stover	1,050	35·1
wheat straw	2,034	25·0	wheat straw	340	4·2
dairy manure	1,143	10·9	dairy manure	684	6·5
		153·0			142·7
Mountain North			*South East*		
forestry wastes	14,326	65·2	forestry wastes	45,529	207·2
maize stover	36	1·2	maize stover	1,902	63·6
wheat straw	2,785	34·2	wheat straw	302	3·7
dairy manure	179	1·7	dairy manure	694	6·6
		102·3			281·1
Mountain South			*North America*		
forestry wastes	10,077	45·9	forestry wastes	489,036	1236·5
maize stover	362	12·1	maize stover	30,026	1004·7
wheat straw	1,409	17·3	wheat straw	35,775	439·6
dairy manure	267	2·5	dairy manure	13,178	125·7
		77·8			2806·5
Plains North					
forestry wastes	1,659	7·5			
maize stover	4,769	159·6			
wheat straw	10,423	128·1			
dairy manure	515	4·9			
		300·1			

Figure 7.16   Biofuels Totals in North America (continued)

large biogas potential of Alaska and the whole of Canada is notable. In the United States the biofuels resource is most significant in the Mountain North and Plains North regions and in New England the Southeast. Smaller but useful quantities of energy could also be obtained in the Lake States, the Corn Belt, and in the Appalachian region.

	Mains gas 1979 PJ	Potential biogas PJ	Biogas as % of mains gas
Alaska	112	40·0	35·7
North	nil	26·4	—
British Columbia	156	55·5	35·6
Canadian Plains	657	205·9	31·3
Eastern Canada	808	156·4	19·4
Maritime Provinces	0·01	12·9	1290·0
Pacific	2192	153·0	7·0
Mountain North	190	102·3	53·8
Mountain South	852	77·8	9·1
Plains North	745	300·1	40·3
Plains South	4538	105·9	2·3
Lake States	1210	222·5	18·4
Corn Belt	4231	667·6	15·8
Delta	1963	67·0	3·4
New England	269	62·9	23·4
Northeast	1737	126·3	7·3
Appalachians	717	142·7	19·9
Southeast	1185	281·1	23·7
North America	21,562	2806·3	13·0

Figure 7.17 Size of the Biofuel Resource in North America

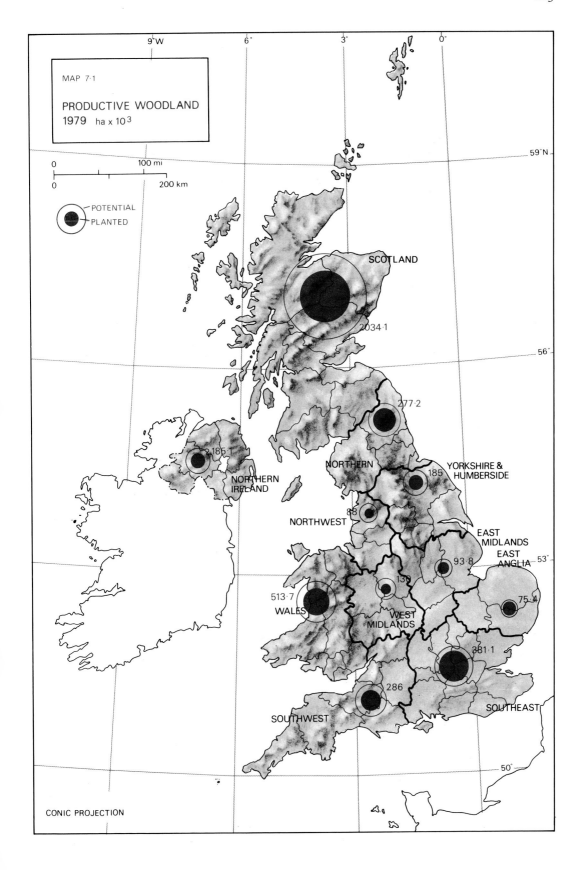

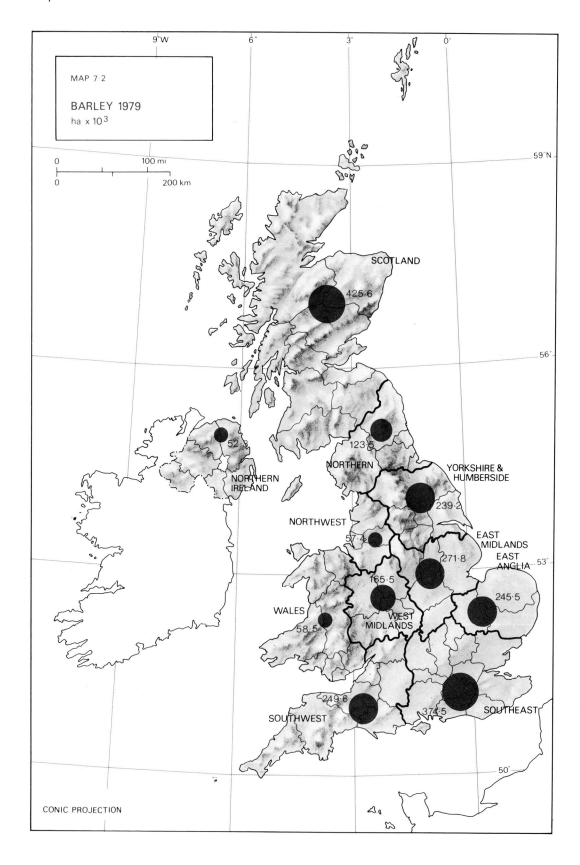

MAP 7.2 BARLEY 1979 ha × 10³

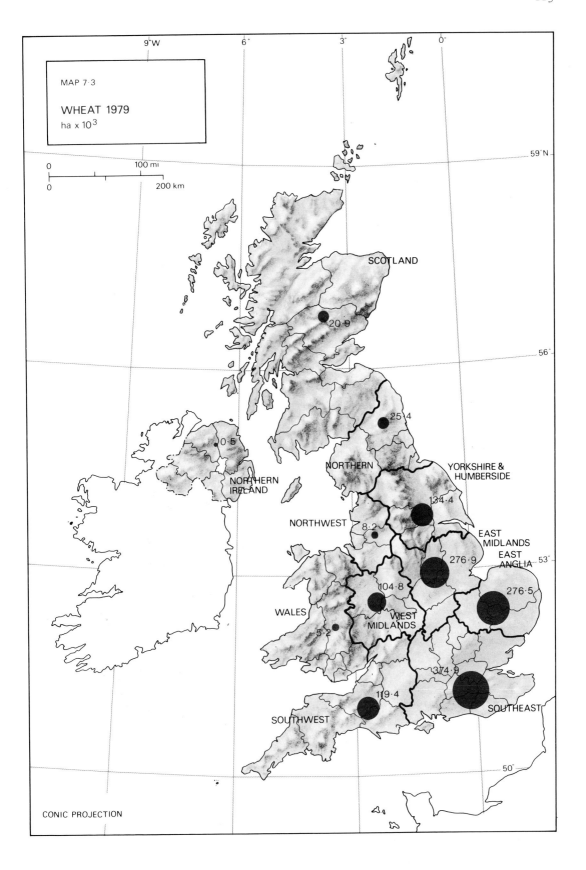

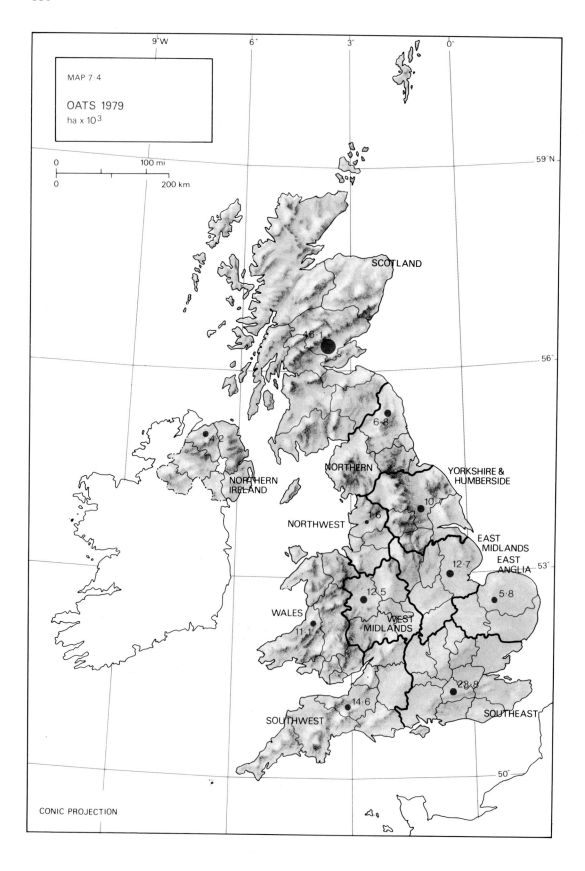

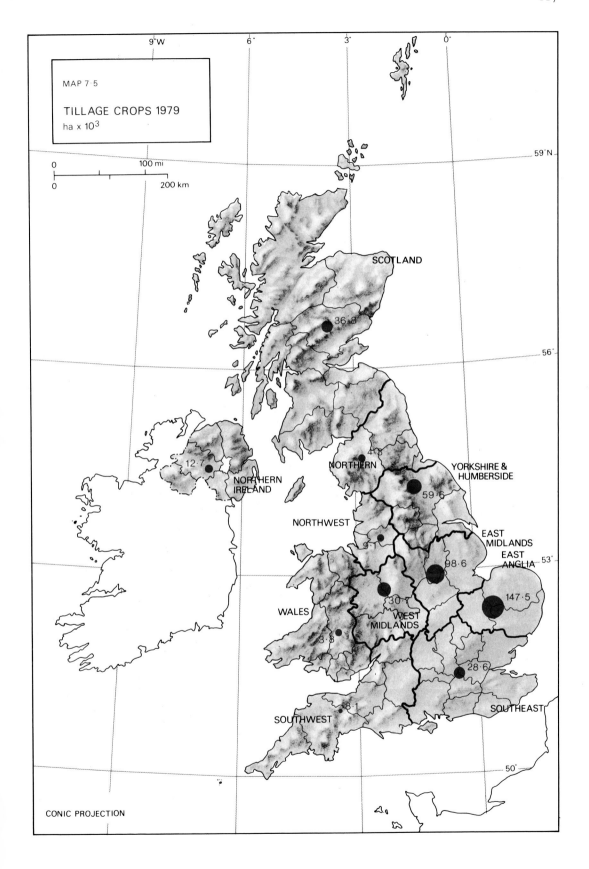

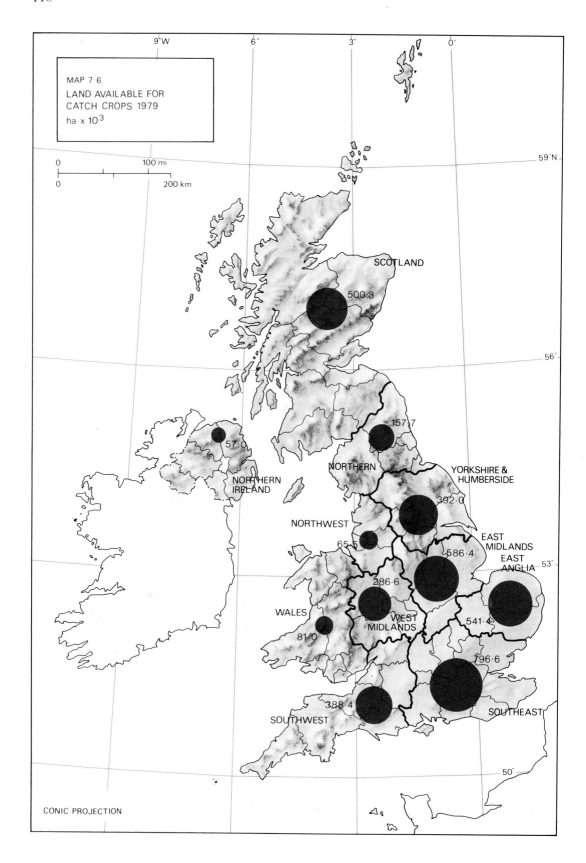

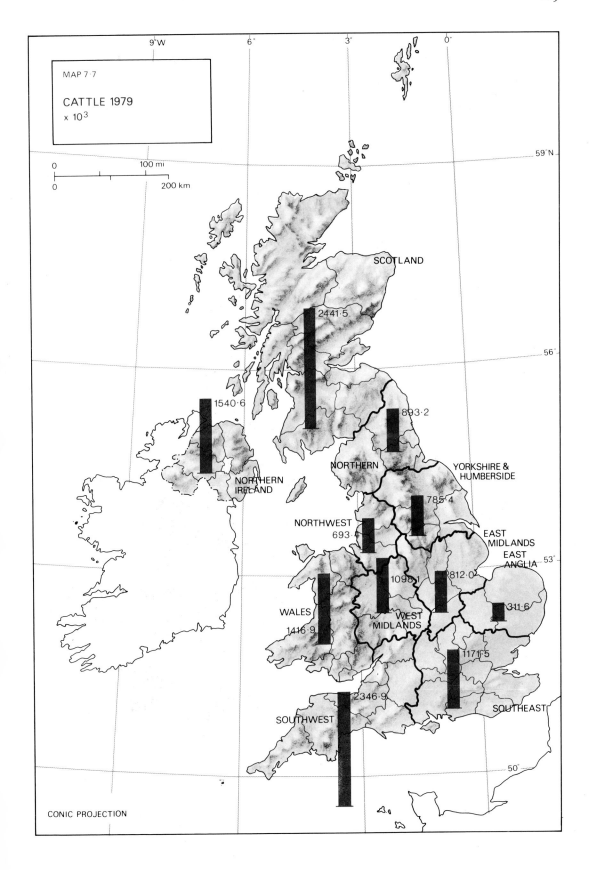

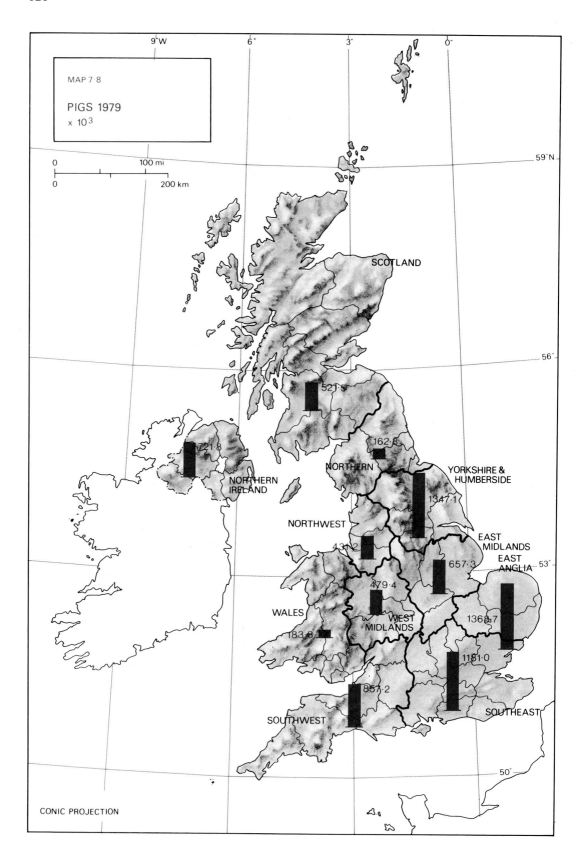

MAP 7.8

PIGS 1979

× 10³

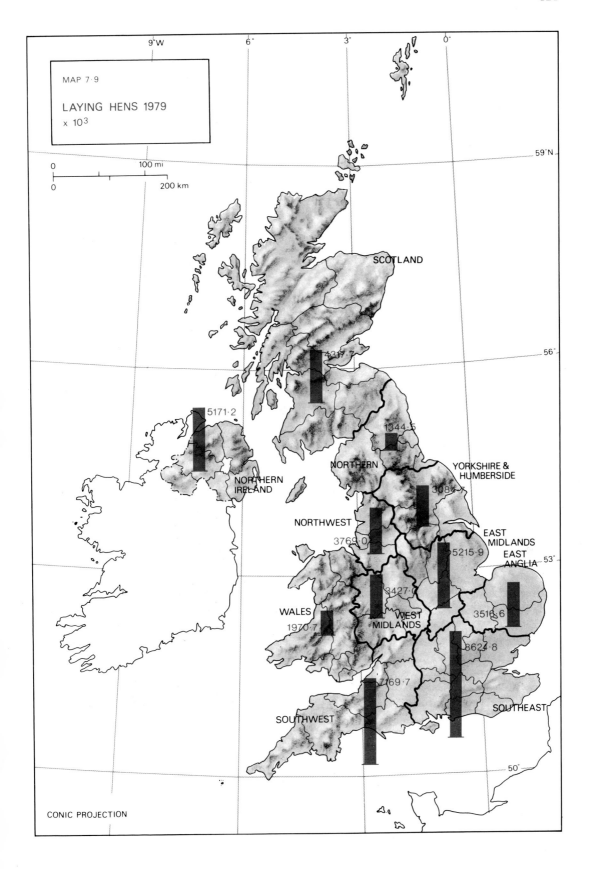

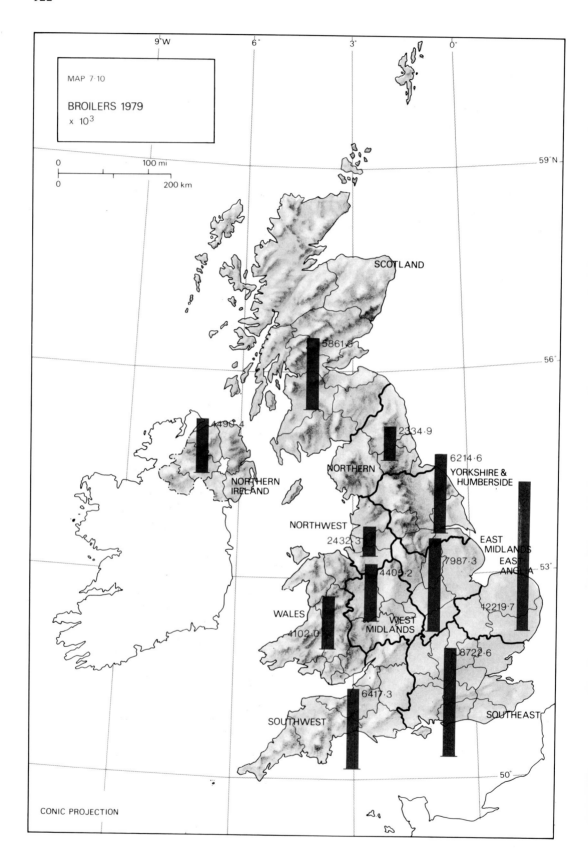

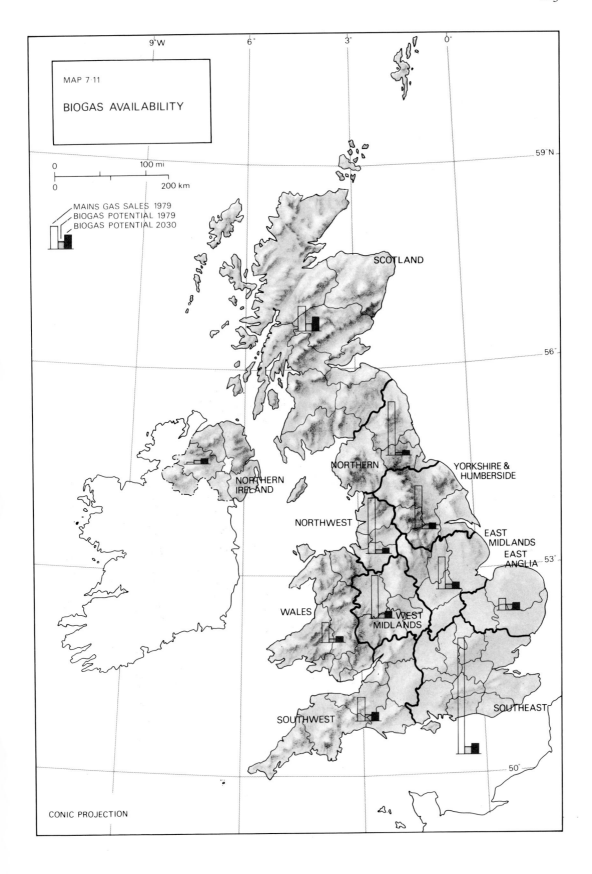

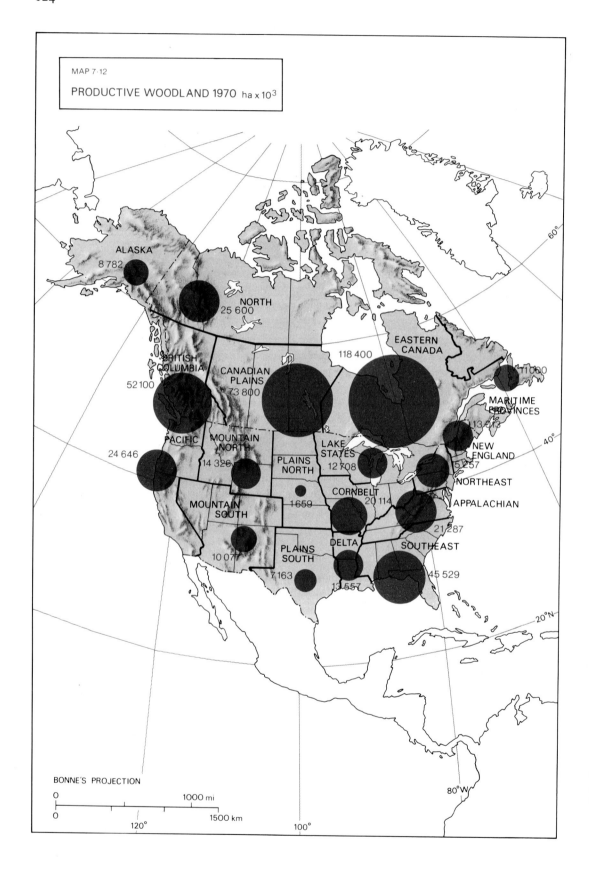

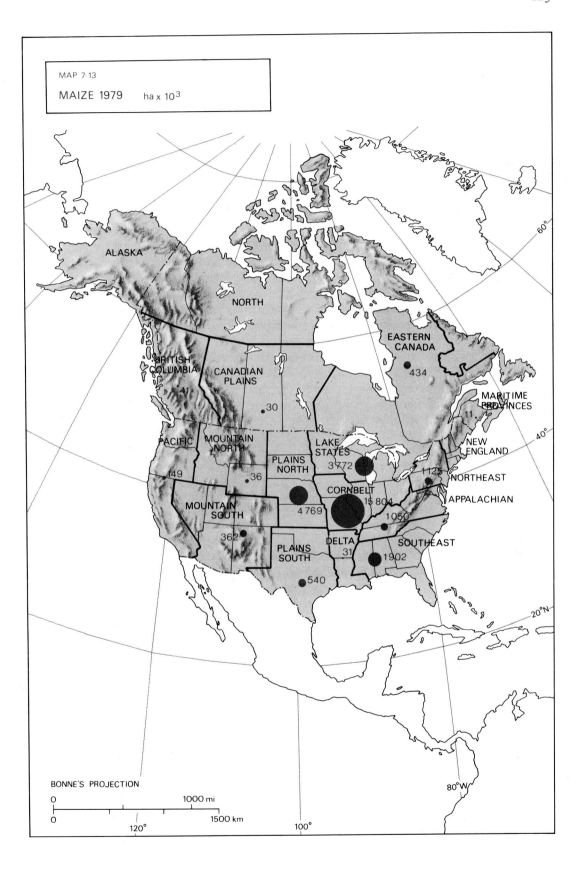

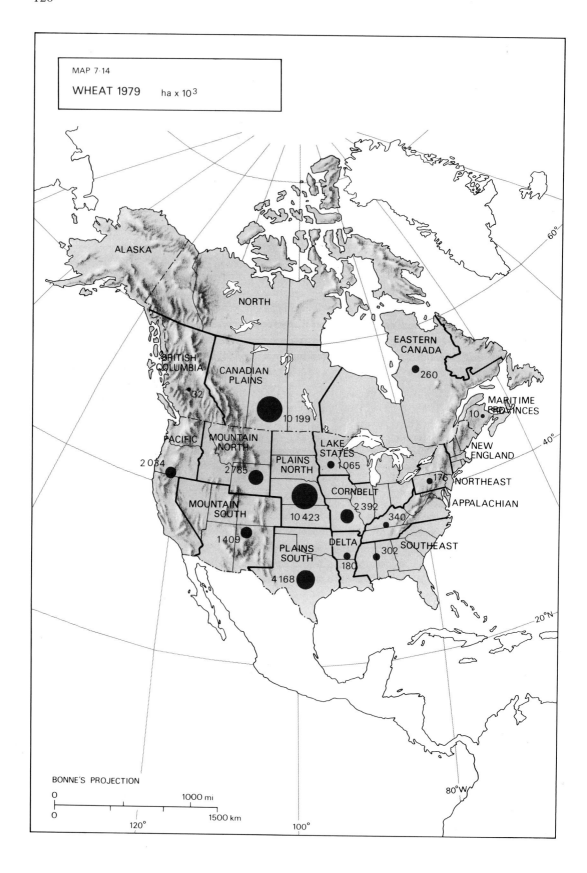

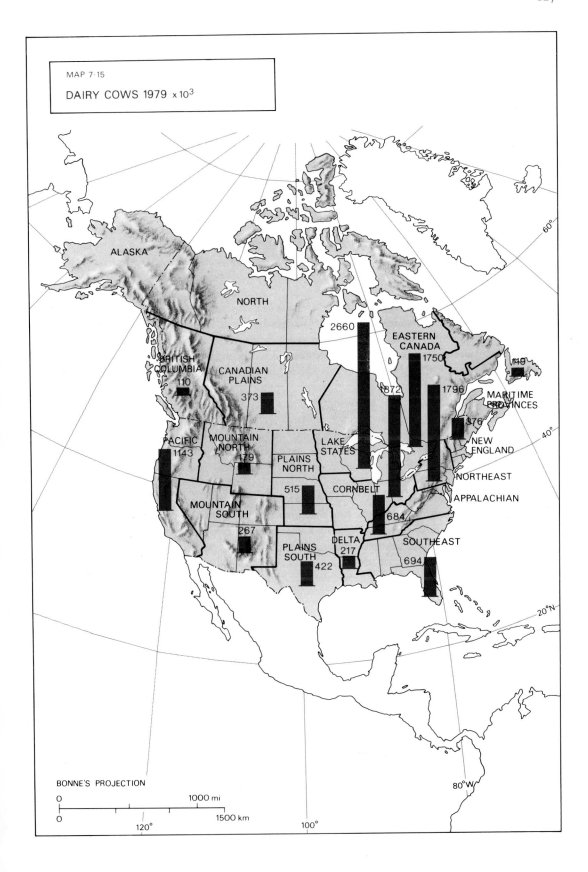

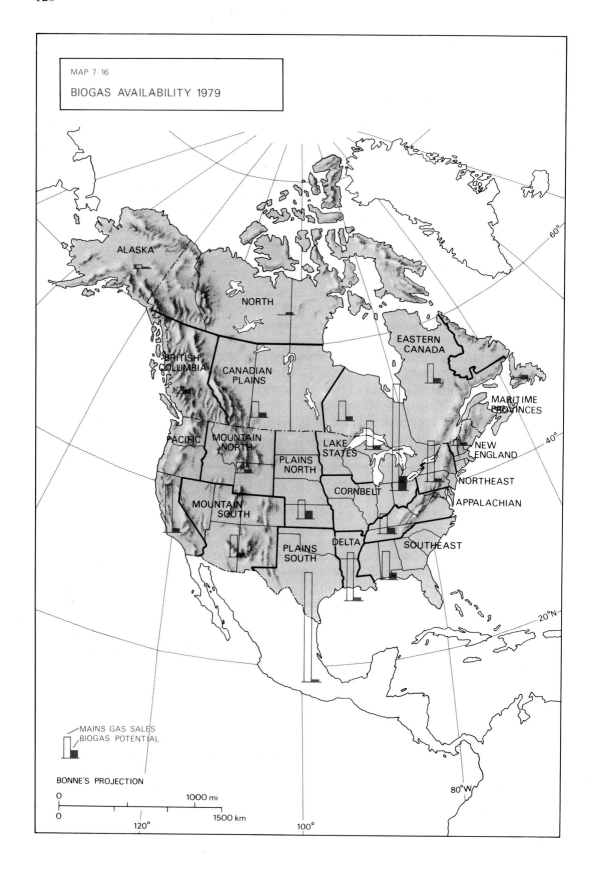

Chapter Eight
Geothermal Energy

It is common knowledge that the temperature of the Earth increases with depth. Readings taken in mines and boreholes provide direct evidence of rising temperatures to depths of as much as 9 km, but because the cost and difficulty of penetrating through rock also increase with depth our knowledge of conditions below about 10 km must be obtained by indirect methods. A remarkably large body of information has been gathered by means of seismic, gravitational, and magnetic studies and Figure 8.1 summarizes some of what is now known about the inner structure of our planet.

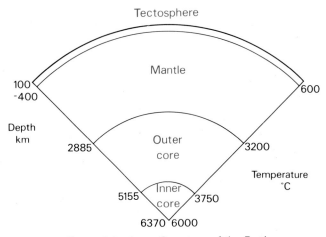

Figure 8.1   Inner Structure of the Earth

    The solid inner core of the Earth, forming only 1·7 per cent of its mass, consists of an alloy of iron and nickel which, despite its high temperature, is prevented from melting by the enormous pressure to which it is subjected by the remaining layers of the globe. Surrounding the inner core is the liquid outer core made up of a mixture of sulphur and iron. Convective activity in the outer core is responsible for the Earth's magnetic field. At a depth of about 2885 km a discontinuity in the Earth's structure occurs at which the rocks forming the mantle meet and contain the liquid material of the outer core. The mantle makes up about two-thirds of the mass of our planet.

    Surrounding the mantle and forming the outermost layer of the Earth lies the composite region known as the tectosphere. Within the tectosphere, whose thickness varies from about 100 km beneath the oceans to some 400 km under the continents, the circulatory processes of plate tectonics take place. Two principal layers in the tectosphere may be distinguished. The deeper layer, the asthenosphere, is composed of the same rocks as the mantle and is in a constant state of slow movement under the influence of convective forces originating deep within the planet. Resting upon it is the lithosphere, or sphere of rock, which is deformed elastically by the shifting motion of the asthenosphere beneath. Mountain building, the wandering of the continents, and the large scale features of the surface of the Earth are the result of movement and deformation in the lithosphere.

In its turn the lithosphere is divided into two zones by the Mohorovicic discontinuity or Moho. Beneath the Moho, which lies between 25 and 90 km below the continents but which rises to within five to 10 km of the surface under the oceans, the rocks of the lithosphere are partly melted and are of low viscosity. This semi-molten material occasionally breaks through to the surface in regions of volcanic activity. Finally, above the Moho, are lighter, rigid, and cooler rocks forming the crust of the Earth. Thus a picture emerges of a multilayer planet built concentrically round a small solid inner core. Upon its spherical crust, which is no thicker in proportion than is the skin of an apple, mankind and all organic creation perform the drama of life.

At the centre of the Earth the inner core has a temperature of about 6000 °C. The temperature at the inner surface of the mantle has fallen to some 3200 °C while the rocks lying at the base of the crust are at about 600 °C. The average surface temperature of the Earth is only a few degrees above freezing. We therefore live on a very hot planet whose temperature falls about a thousand-fold between the centre and the surface of the crust. Most of the fall takes place in the mantle. The thermal condition of the interior of the planet is kept constant by the heat produced by the radioactive decay of naturally occurring isotopes of uranium, thorium, and potassium.

The rocks of the mantle and crust are poor conductors of heat and function in the manner of an insulating jacket surrounding a tank of very hot water. Nevertheless, some heat escapes from the Earth's interior and is lost to space. The world average outward heat flow of 60 mW m^{-2} is not perceptible by our senses but when summed over the whole surface of the globe it amounts to the immense energy flow of 30 TW. This is about 17 times the capacity of installed electric power plant in the world, but because the flow is so diffuse there is little hope of tapping it for human purposes.

The heat reserves within the Earth are so gigantic as to be, for human purposes, unlimited. However, until it is possible to sink a borehole as deep as 30 km and thus to tap the heat reserves of the rocks below the Moho, our attention must be confined to the more limited but still very large heat content of the solid rocks of the Earth's crust. The deepest borehole ever sunk, Baden No. 1 in the United States, reached 9·17 km while the deepest hole in the United Kingdom was made at Seal Sands on Teesside in 1975 when a depth of 4·12 km was attained. Drilling operations of this scale are difficult and extremely expensive. In this study, therefore, only the top 6 km of the Earth's crust are taken into consideration.

The rocks of which the crust is composed fall into two main types. Those whose origin was igneous, such as basalt or granite, are largely impermeable to water. Although igneous rocks will entrap water in the cracks and fissures occurring in their mass, the amount of water is usually small and it is useful to regard them for energy purposes as dry and, at depth, hot. The water naturally present in these rocks will play little part in the exploitation of their heat reserves.

Sedimentary rocks such as limestones and sandstones, however, will allow water to pass into and through their porous structures. These rocks absorb water as if they were huge sponges and are saturated with water often to very great depths. Other types of sedimentary rocks, notably clays and mudstones, are impervious to water. A succession of sedimentary rocks will often contain interleaved layers of wet porous and relatively dry impervious rocks. In this way conditions are created in which large bodies of hot or warm groundwater are accumulated within sedimentary rock formations. These reservoirs are known as aquifers.

Sedimentary rocks, because of the process of deposition by which they were formed, will occupy any basins occurring in underlying older impervious formations. Some sedimentary basins in the United Kingdom are more than 3 km deep while in southern Texas the bottom of the North Gulf of Mexico Basin lies more than 12 km below the surface. Aquifers occurring in the deep sedimentary rocks within these basins are likely, because of the high temperatures expected at depth, to be good sources of earth heat. The geothermal reserves of an aquifer are tapped by bringing to the surface some of its hot groundwater.

As will be seen from the maps accompanying this chapter, the pattern of surface heat flow does not correspond very closely to geological structure either in the United Kingdom or in North America. Since the detail of the geology of even these well-studied parts of the Earth's crust is poorly understood, and in particular the arrangement of faults and fissures at depth are unknown, it is always possible to explain away an apparent anomaly by postulating a migration of groundwater. Alternatively, our ignorance leaves room for speculation about bodies of hot magma rising to near the surface at places where the crust may be thin. These are undoubtedly the correct kinds of explanation for many of the observed facts, but it should be recognized that much more information is needed before all geothermal anomalies can be accounted for in a quantitative way.

THE QUANTIFICATION OF GEOTHERMAL ENERGY

The amount of heat contained in a mass of rock is a function of the thermal capacity of the material and of its temperature. The density and specific heat of a formation can be found experimentally and it is a simple matter to calculate the heat capacity of a volume of rock using the ascertained values. However, the fact that the temperature of a rock formation increases with depth, causing a corresponding increase in thermal content, calls for the introduction of an important new parameter known as the geothermal temperature gradient. Figures 8.2 and 8.3 are graphs illustrating this relationship as found in boreholes in the Cornish granite and in a part of the Wessex sedimentary basin. The location of these two boreholes is shown on Map 8.1.

Figure 8.2 describing the geothermal gradient of a mass of granite at Rosemanowes in Cornwall shows that its temperature increases evenly with depth at a slightly increasing rate averaging about $32\,°\mathrm{C\,km^{-1}}$. The straightness of the line follows from the fact that the rock is relatively dry, free from faults, and homogeneous in character. The sedimentary rocks through which the borehole made at Winterborne Kingston penetrate are made up of a complex sequence of

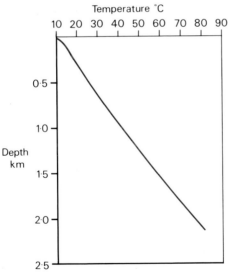

Figure 8.2 Geothermal Gradient in Borehole RH 11 at Rosemanowes

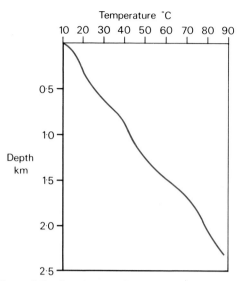

Figure 8.3   Geothermal Gradient at Winterborne Kingston

limestones, sandstones, and shales, each with differing thermal properties. The wavy line shows that the average gradient of $30\,°C\,km^{-1}$ conceals large variations in the gradient at various depths.

It is customary to describe geothermal temperature gradients by means of a three-part classification.

1 Hyperthermal regions where the gradient exceeds $80\,°C\,km^{-1}$.
2 Semi-thermal regions where the gradient lies between 40 and $80\,°C\,km^{-1}$.
3 Normal regions where the gradient is less than $40\,°C\,km^{-1}$.

Figure 8.4 illustrates these categories. Only in a few small areas of the British Isles do geothermal gradients exceed the normal range (Burley and Edmunds, 1978), but hyperthermal gradients are quite commonly encountered in the western parts of North America.

The usefulness of a supply of heat is dependent not only upon the quantity available but also upon its temperature, or as it is often expressed upon its grade. It will be seen from Figure 1.2 that if useful steam is to be raised from geothermal sources then rocks at a temperature of $200\,°C$ must be reached. Many industrial processes call for heat at around $130\,°C$ while $60\,°C$ can be adequate for space heating in buildings. Consequently, the quantity of useful heat which is available from the Earth will vary according to the purpose for which it is needed. In effect, the higher the temperature required the deeper will be the surface layer of rock which must be excluded from the quantification of available heat as being too cool.

The average depth at which rock of a given temperature will be encountered is given by:

$$Z = T/G\,\text{km}$$

where T = temperature required (°C)
G = average temperature gradient (°C km^{-1})

Figure 8.5 shows the depths which must be reached if rocks of the three useful temperatures described above are to be found. The range of temperature gradients included on the abscissa covers the conditions to be found in normal and semi-thermal regions.

It is evident that boreholes sunk for the purpose of obtaining heat for space heating need to be

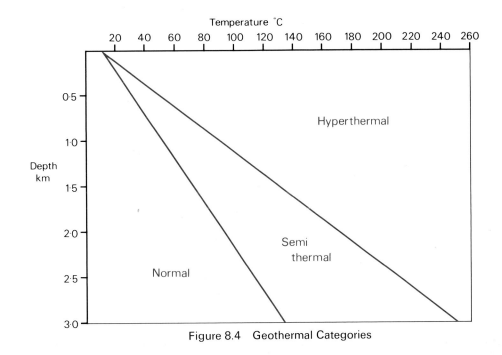

Figure 8.4 Geothermal Categories

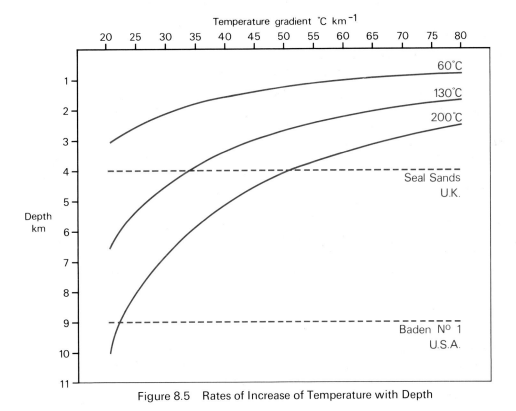

Figure 8.5 Rates of Increase of Temperature with Depth

only about one-third the depth of those intended to raise steam. It should be noted also that the 60 °C curve is flatter than those for 130 or 200 °C. Consequently the depth to which a space heating well must be taken is less affected by the thermal gradient encountered than are the depths required for process heating or steam raising wells.

HEAT CONTENT OF HOT DRY ROCKS

Granite has a density of 2.7 g cm^{-3} and specific heat of $0.82 \text{ J g}^{-1} \text{°C}^{-1}$. With these values it can be shown (Garnish, 1976) that the heat content, Q, of a mass of granite may be found from the following formula:

$$Q = 1.1 \, G(Z - Z_0)^2 \times 10^{15} \text{ J km}^{-2}$$

where G = geothermal temperature gradient (°C km^{-1})
Z = lower depth (km)
Z_0 = upper depth (km)

It will be seen that the energy content obtained from this formula is that of an area of 1 km² of granite lying between the two specified depths. The dimension Z is the depth to the bottom of the mass of rock in question while the upper depth, Z_0, is determined by the minimum temperature at which the heat energy is required. The heat contained in the shallower and cooler rock above this depth is excluded from the result.

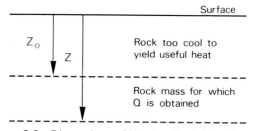

Figure 8.6   Dimensions of Hot Dry Rock Heat Source

Figure 8.7 shows the curves obtained when the equation is entered with geothermal gradients characteristic of normal and semi-thermal regions. The three sets of curves give values for space heating, process heating, and steam raising purposes. Against each curve is placed the depth Z forming the lower boundary of the rock mass.

Granites and other igneous rocks are too dry to enable their little naturally-occurring water to function as the medium for heat extraction. However, recent work at the Los Alamos Scientific Laboratory in New Mexico has shown that techniques of hydrofracturing can produce a porous structure at depth in even very hard rocks. The large surface area created by the fracture can transfer heat to injected water very efficiently. A pumped circulation system can then be used to carry the heat to the surface for use.

HEAT CONTENT OF AQUIFERS

Many sedimentary rocks are, unlike granite, porous and capable of holding water in the interstices of their structure. This property may be defined in terms of a percentage porosity. A porosity of 10 per cent, for instance, describes a formation composed of 90 per cent rock and 10 per cent water by weight.

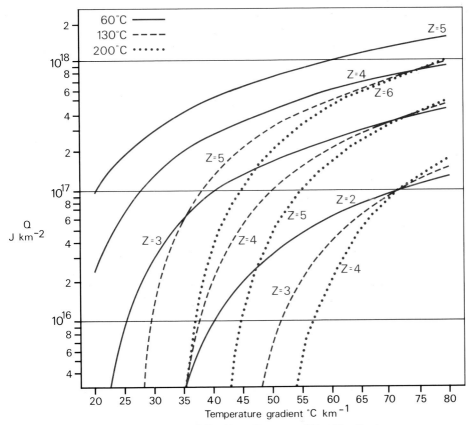

Figure 8.7 Heat Content of Hot Dry Rocks

Both sandstone and limestone may be taken to have a density of $2 \cdot 4\,\mathrm{g\,cm^{-3}}$ and a specific heat of $0 \cdot 84\,\mathrm{J\,g^{-1}\,°C^{-1}}$. The specific heat of water is $4 \cdot 18\,\mathrm{J\,g^{-1}\,°C^{-1}}$ and its density is, of course, unity. Since the specific heats of rock and water are so different the thermal properties of an aquifer will vary with the amount of water entrained. For this reason the following formula, from which the quantity of heat contained in an aquifer can be calculated, must be entered with a figure for the porosity of the rock of which the aquifer is formed.

$$Q = T(Z - Z_0)(2 \cdot 02 + 2 \cdot 16p) \times 10^{15}\,\mathrm{J\,km^{-2}}$$

where T = average temperature (°C)
Z = lower depth (km)
Z_0 = upper depth (km)
p = porosity expressed as a decimal percentage
 sedimentary rocks vary between 5 per cent and 25 per cent porosity

The interpretation of Z and Z_0 must take account of two characteristics of sedimentary rocks. The strata in a sedimentary basin which are capable of bearing aquifers are often overlain by unconsolidated rocks, frequently of glacial origin, which are too porous to contain groundwater. If the unconsolidated formation is very thick, the upper boundary of an aquifer may be below the depth at which, from a consideration of thermal gradient only, useful temperatures might be expected to be found. Secondly, aquifers do not occur throughout the depth of sedimentary formations. Trial boreholes are therefore needed to establish values of Z and Z_0 as well as T and p in the above equation.

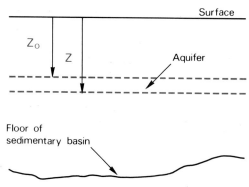

Figure 8.8 Dimensions of an Aquifer

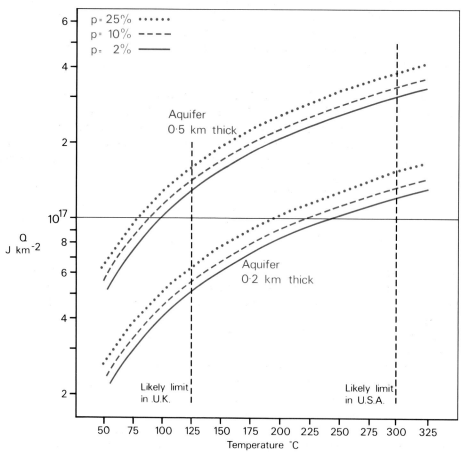

Figure 8.9 Heat Content of an Aquifer

When porosities of 2 per cent, 10 per cent, and 25 per cent are entered into the equation the results shown in Figure 8.9 are obtained for aquifers 0.2 and 0.5 km thick. The curves show a higher thermal content than would a similar volume of granite because the specific heat of the water entrained in the aquifer is nearly five times that of granite. It is unlikely that the temperature of any aquifer in the British Isles will much exceed the boiling point of water because the deepest sedimentary basin, the Wessex Basin, does not reach 3.5 km. In the United States, however, the floor of the Gulf of Mexico Basin is more than 12 km beneath the coast of Texas. A geothermal gradient of $25\,°C\,km^{-1}$ is sufficient to produce a temperature of $300\,°C$ at this depth.

THE MAPS

High rates of surface heat flow are generally an indication of high temperatures at depth. The heat may reach the surface from a hot aquifer below. Alternatively, it may originate in rocks containing concentrations of radioactive material or from an intrusion of hot magma. Whatever its origin, the flow of heat may be impeded by overlying rock formations with a high thermal resistance and it can be displaced laterally by movements of hot groundwater. For these reasons the correspondence between geological structure and heat flow rate is not a direct one. Nevertheless, regions in which high heat flows coincide with promising rock formations invite geothermal exploration and it is useful to know where these areas are.

Map 8.2 shows that most parts of the British Isles experience a rate of terrestrial heat flow below the world average of $60\,mW\,m^{-2}$. The modest heat flux results from the fact that the islands lie far from an active edge of a continental plate and have enjoyed a long period of tectonic tranquility. But there are some areas where the flux is relatively high and it would be useful to know what correspondence exists between heat flow rates and the geological structure of the country.

Formations of igneous rocks are, in the British Isles, confined to northern and western regions. A group of tertiary volcanoes have left a record of their activity in the Inner Hebrides on Skye, Rhum, Ardnamurchan, Mull, on Arran, and in Ulster. It is likely that the zone of higher heat flow centred on Lochaber is associated with these events, and it is also highly probable that other areas of high flux will be found when the Inner Hebrides and Northern Ireland are surveyed for their thermal resources.

The older granite intrusions of Galloway, which are of Ordovician age, seem to have no effect upon surface heat flow. The land to the north and south of Morecambe Bay exhibits higher than average heat flow. This may be attributable to the Lake District granites or, alternatively, it may be the result of the fact that the Cheshire sedimentary basin extends as far north as St Bees Head.

The northwestern part of the county of Durham and the adjacent parts of Northumberland and Cumbria are underlain by the Weardale granite intrusion. It is of Silurian origin and although it nowhere reaches the surface it appears to have a profound effect upon the terrestrial heat flow in the vicinity. The flux exceeds $90\,mW\,m^{-2}$ over a considerable area. This part of the country would probably repay a detailed geothermal resource survey.

South Devon and almost the whole of Cornwall is founded upon a huge granite mass of Hercynian age which extends westward past the Scilly Isles an unknown distance into the Atlantic Ocean. Above average heat flows are associated with this formation reaching $120\,mW\,m^{-2}$ at Hensbarrow, Carnmenellis, and Lands End. These are the highest terrestrial heat flows in the British Isles and indicate the existence of substantial geothermal resources in the southwest.

It is surprising that the Lincolnshire sedimentary basin, which near the coast exceeds 2 km in depth, underlies a part of the country where the terrestrial flux is low. Nottinghamshire, however, whose geological structure would not appear to promise resources of heat, lies at the centre of an area of heat flux as high as $80\,mW\,m^{-2}$. The northeastern part of this warm patch embraces the known thermal districts of Matlock and Buxton.

The large area within the $40\,mW\,m^{-2}$ experiences the lowest heat flux in the British Isles. This is a remarkable anomaly, for it lies over the Cheshire–Shropshire and Worcestershire basins where geothermal resources might be expected. No explanation able to account for these facts is readily available. The thermal flux is low over the whole of Wales.

It appears that the thermally most favourable sedimentary formation in Britain is to be found in the western portion of the Wessex basin. The $80\,mW\,m^{-2}$ isopleth extends as far east as Poole and it seems very possible that the conurbation of Poole–Bournemouth could obtain much of its requirement for low grade heat from the Earth. The average or below average heat flux experienced by the counties of Hampshire and West Sussex, which also overlie the Wessex basin, is presumably to be accounted for by movement of groundwater.

THE SIZE OF THE RESOURCE

It has been shown that the thermal energy content of a mass of rock, sedimentary or igneous, can be ascertained if certain facts about it are known. The facts required of hot dry rocks are the superficial extent of the formation, its depth, and the geothermal temperature gradient. Measurements of the extent, depth, thickness, and temperature gradient of an aquifer lying within a mass of sedimentary rock are needed to calculate its heat content. It is unfortunate that even in the British Isles, whose geology has been intensively studied for two centuries, too little data exists for a comprehensive and quantitative assessment of the geothermal energy resource of the nation to be made.

Map 8.1 shows the position and extent of the chief igneous formations in Britain as they reveal themselves at the surface, or in the case of the Weardale granite very near the surface. Neither their depth nor their lateral extent at depth are known with any accuracy. It is very likely, for instance, that all the granite areas in southwest England are part of a large batholith but a thorough survey would be needed to confirm this supposition and delineate its boundaries. Similarly, a common foundation may underlie and unite the igneous formations of the Inner Hebrides.

The depth and extent of the sedimentary basins of the British Isles are well known, but little has been discovered about the size and temperature of any aquifers they may contain. Here too much survey work is required if the facts are to be established. Geological investigations are laborious and, because of the high cost of drilling, very expensive. They have therefore always been undertaken sparingly and only in response to a clearly understood economic need. Appreciable amounts of earth heat exist in Britain, but until money is made available and much more data obtained it is impossible to discover what is the size of the geothermal energy resource in these islands.

THE GEOTHERMAL ENERGY REGIME OF NORTH AMERICA

The most valuable geothermal fields, from the point of view of electricity production, are those which yield naturally-occurring steam rather than hot or warm water. Unfortunately, fields of this kind are rare even in the hyperthermal regions of western North America. Only three vapour-dominated systems are known and two, the Lassen system in California and the Mud Volcano area at Yellowstone, are not available for exploitation because they are within national parks. The third, the Geysers system north of San Francisco, has been used to generate electricity on a commercial scale since 1960. The resource at The Geysers is estimated to be capable of supplying about 15 PJ annually for 100 years. Resource estimates given in this section are those furnished by the United States Geological Survey (Muffler, 1979).

The estimates of the life of a resource are based upon the assumption that no thermal recharging of the system takes place during the period of time when heat is being extracted. There is good reason to believe that many geothermal systems are, in fact, capable of gaining heat from deeper layers of the Earth's crust at a significant rate (Muffler and Cataldi, 1978). However, the extent to which this process can lengthen the useful life of the resource cannot be predicted quantitatively. It has been ignored in the present evaluation and the figures presented should therefore be regarded as conservative.

Hot water geothermal fields are to be found in many parts of western North America. The quantity of electricity that can be generated from those fields producing water at more than 150 °C is much larger than that obtainable from steam fields. This resource is estimated to be able to contribute, from known fields, 198 PJ of electricity to the North American economy for a century. Most but not all the fields delivering water at 150 °C or above are within the 100 mW m^{-2} isopleth on Map 8.4.

The electricity production from steam and hot water fields in North America in 1979 totalled 19 PJ. However, only a fraction of the known resource is at present exploited, and many more geothermal fields remain to be discovered. An annual total of 1183 PJ is expected to be available when further exploration has identified the whole of the accessible resource. The immense thermal resources of hot dry rocks in North America are not taken into account when arriving at this estimate.

Although it cannot be employed for the purpose of raising steam a wide variety of uses can be found for geothermal heat at temperatures between 90 and 150 °C. An average annual value, when the full size of the resource is known, of the heat likely to be available in North America at these temperatures for the period of a century is the enormous quantity of 15 EJ. Energy of this grade would be used chiefly for space heating in buildings and as a source of industrial process heat. But the location of the heat source and the heat load is very important in the case of medium temperature applications because of the difficulty and expense of transporting heat in this form. It is impossible to say how many of these interconnections will, in the future, be short enough to be effective and for this reason no attempt is made in this atlas to quantify further the medium temperature geothermal resource.

The amount of Earth heat below 90 °C that could be obtained in North America is not known, and very extensive surveys would be needed before the occurrence of this type of energy could be mapped statistically. In the meantime two avenues of enquiry can furnish a rough impression of the geography of the low temperature geothermal resource. Firstly, the United States Geological Survey lists four favourable areas for discovery in Alaska, five in Hawaii, 135 in the western part of the United States, and 40 in central and eastern regions. These records are based upon very imprecise data and there is little doubt that a thorough survey would be rewarded by the discovery of many more useful low temperature heat sources.

Secondly, a comparison of Map 8.3, showing the principal sedimentary basins of North America, and Map 8.4 of terrestrial heat flow rates can suggest those parts of the continent in which heat-bearing aquifers might be found. The correlation between geological structure and heat flow rates is no more exact in North America than it is in the British Isles, but some correspondences should be pointed out.

The North Alaska Coastal Basin, which slopes inwards towards the mountains of the Brooks Range, is, together with the Peel Basin beneath the mouth of the Mackenzie River, an area of higher than average heat flow. Even higher values, of over 120 mW m^{-2} are recorded on Cornwallis Island in the Canadian arctic archipelago (Garland and Lennox, 1962). The high heat flow rate in southeastern Alaska, which presumably extends south into the Canadian Rocky Mountains, must be the result of tectonic movement between the North American and Pacific plates. It is unfortunate that no more detailed description of western Canada can be offered, for a comprehensive account of terrestrial heat flow in all parts of Canada has not yet been attempted. But when this task is carried out it will probably be found that the heat flow isopleths in the western part of the country are chiefly a reflection of the presence of hot intrusive rocks. The Alberta Basin does not appear to influence the rate of heat flow at the surface.

The same cannot be said of the Williston Basin under North Dakota, where the nearby $80\,mW\,m^{-2}$ isopleth may well reflect its presence. A similar near but not exact correspondence exists between the Anadarko Basin and the $80\,mW\,m^{-2}$ isopleth in northern Texas. In the most southerly region of Texas, however, high rates of heat flow are found directly above the very deep North Gulf of Mexico Basin. This large basin may well contain extensive hot aquifers. The low terrestrial heat flows in Pennsylvania, West Virginia, Kentucky, and Tennessee do not lead to the same conclusion about the Allegheny Basin. Similarly, the Michigan basin is not reflected in the pattern of surface heat flow. The Black Warrior Basin beneath north Alabama may, however, divert the $60\,mW\,m^{-2}$ isopleth westward in nearby northern Georgia.

It would appear that the North Gulf of Mexico Basin, and to a lesser extent the Williston, Anadarko, and Black Warrior basins are the areas of North America most likely to contain useful geothermal aquifers.

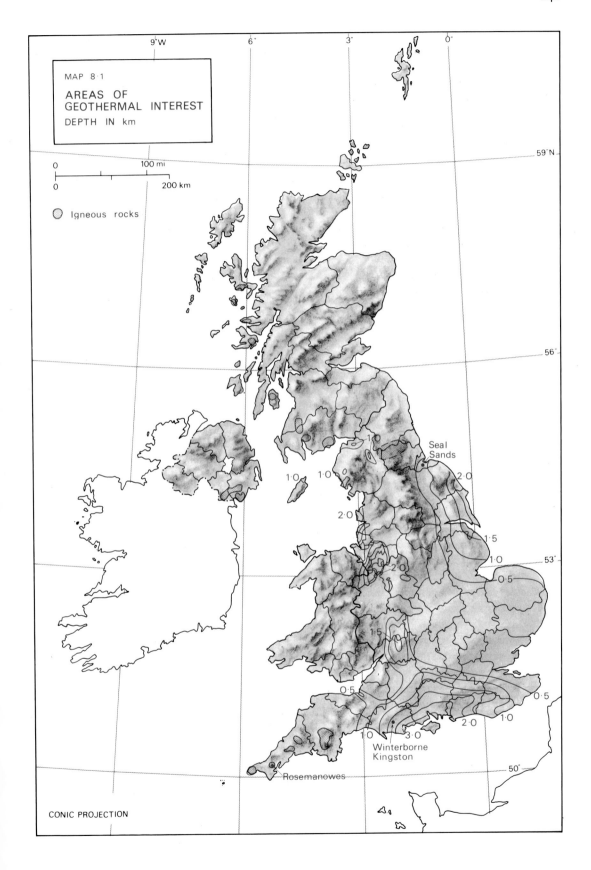

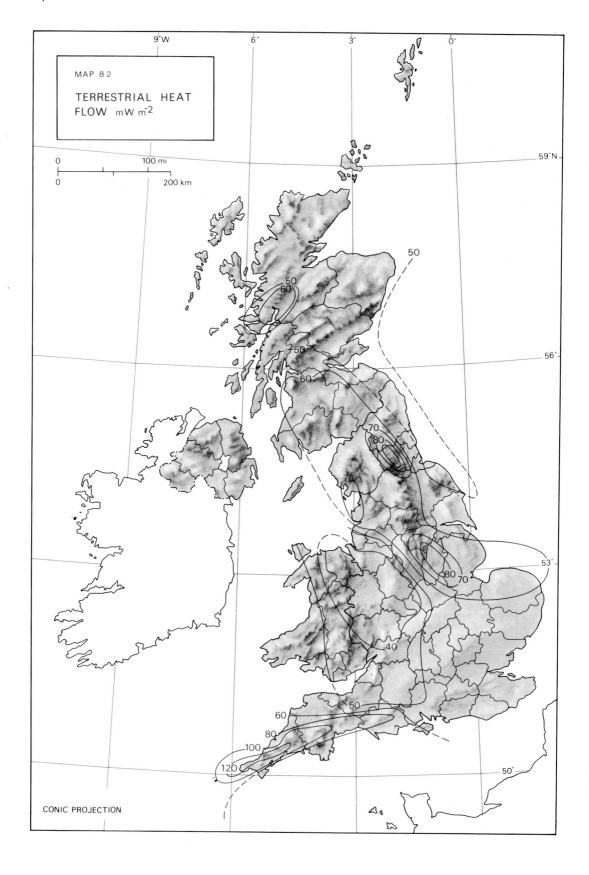

MAP 8.2

TERRESTRIAL HEAT FLOW mW m^{-2}

CONIC PROJECTION

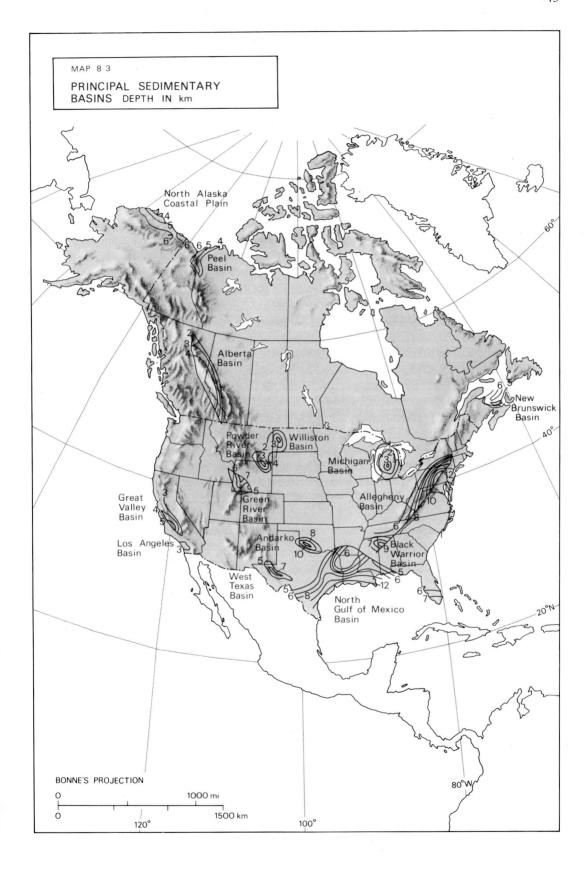

MAP 8·3
PRINCIPAL SEDIMENTARY BASINS DEPTH IN km

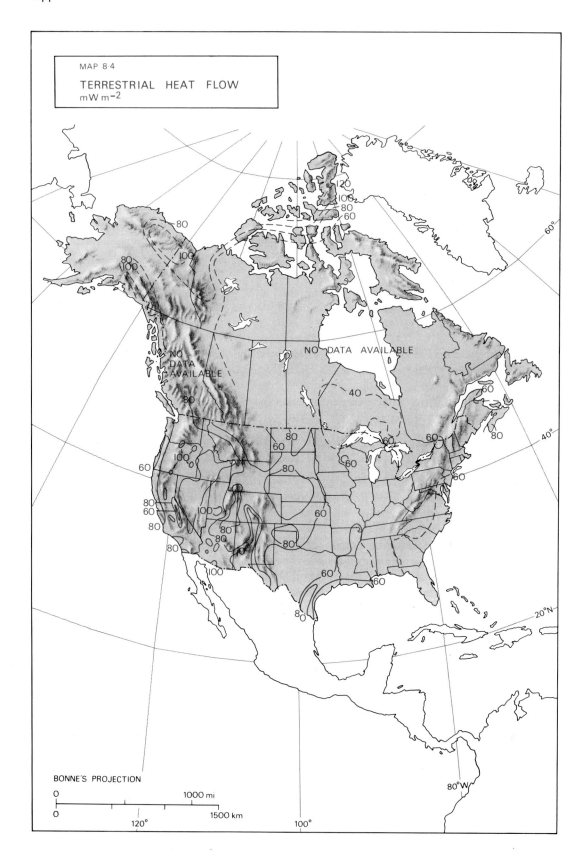

Chapter Nine
Tidal Energy

THE NATURE OF THE TIDES

If the Earth were a perfectly smooth sphere poised in space and unaffected by any external gravitational field the water of which the oceans are composed would cover the planet uniformly to a depth of 2650 m. Our real Earth, however, has a wrinkled surface composed of mountains, plains, and ocean basins. Furthermore, its motion through space is governed, as is that of all the universe, by the force of gravitation. One effect of the gravitational fields in which our planet exists is the regular perturbations of the seas and oceans known as the tides.

The surface of the waters of the Earth rises and falls once or twice a day in response to gravitational forces exerted upon them by our nearest neighbours in space, the Sun and the Moon. Although the Sun is 2.7×10^7 times as massive as the moon it is 389 times as far away from the Earth. Since the tide raising force exerted by an astronomical body varies directly as its mass but inversely as the cube of its distance, it follows that the Sun's tide raising force is only 0.46 times that of the Moon. It is for this reason that the motion of the Moon has from earliest times been recognized as the controlling factor in the timing and behaviour of the tides.

The mechanism by which the Moon raises a tide on Earth can be explained by reference to Figures 9.1, 9.2, and 9.3. Figure 9.1 shows that the radius of the Earth is about 1/60th the distance between its centre and that of the Moon. Since the radius of the Earth is a significant proportion of the distance between the two bodies it follows that the force exerted by the Moon on the solid Earth at its centre is less than that exerted upon a parcel of water under the Moon at U, where the Moon is closer, and greater than the force on a similar but more distant parcel opposite the Moon at U'. The relative strengths of these forces are represented at an exaggerated scale in Figure 9.2. The effect of these three forces may be summarized by saying that at U' the

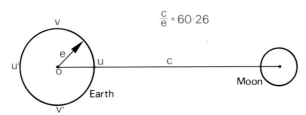

Figure 9.1   The Earth–Moon System

Figure 9.2   Tide Raising Forces

Earth is being attracted away from the water by the gravitational attraction of the Moon, whereas the water is being pulled away from the Earth under the Moon at U. Consequently the Earth's envelope of water is distorted by the force of the Moon into an ellipsoid in the manner shown in Figure 9.3. The points V and V' are separated from the Earth–Moon axis by 90 degrees. They and the centre of the Earth are equidistant from the Moon and its gravitational attraction is therefore the same at all three points.

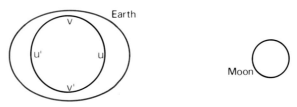

Figure 9.3 Ocean Tides

The distortion of the Earth's envelope of water is further exaggerated by the effect of the revolution of the Earth–Moon system about its mutual centre of mass, a point lying about 1600 km below the surface of the Earth. This whirling motion produces an acceleration in the plane of the Earth–Moon axis which renders the ellipsoidal shape of the water more pronounced. The same gravitational and centrifugal forces are responsible for small but detectable tides in the solid Earth, and indeed in the atmosphere, but these phenomena are of course not significant as an energy resource for man.

It is well known that the tides as they are experienced display not only a periodic elevation and depression of the surface of the ocean but also that the water itself flows horizontally back and forth in the same rhythm. These flows of water are known in British terminology as tidal streams and in North American usage as tidal currents. They are the product of the parallelogram of forces shown in Figure 9.4. Although the differential forces at V and V' are zero there is a slight compressive force exerted at these points by the fact that the Moon's gravitational force acts at a small angle to the Earth–Moon axis. Arrows at V and V' represent this effect. Gravitational forces act perpendicularly to the surface at U and U'. At intermediate latitudes the tide raising forces set up parallelograms of forces whose resultants act toward the points on the Earth–Moon axis where high tides are formed. These resultants, known collectively as the tractive force, transport the water towards places of high tide and are responsible for the ebb and flow of tidal

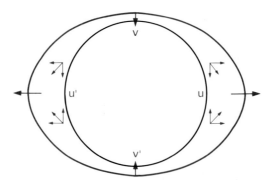

Figure 9.4 The Tidal Tractive Force

streams. During the Earth's daily rotation upon its axis a point on the surface will encounter two raised and two depressed bodies of water and so experience a typical semi-diurnal tidal regime. In fact, because of the relative movement of the Sun, Moon, and Earth an average of 1·94 tidal cycles or 3·87 tidal oscillations occur every 24 hours.

The gravitational force of the Sun produces a similar though less marked effect upon the behaviour of the seas and oceans. When the Sun and Moon are in alignment, at full and new Moon, the two influences augment one another and tides with a large range known as spring tides occur. When, however, the two bodies are at right angles in space their gravitational interaction reduces the tidal range to produce neap tides. Successive spring and neap tides occur at roughly two-week intervals.

Tides, although the result of astronomical forces, are not solely the product of gravitation. A body of water which is in rotation, like the waters of the spinning Earth, will move in such a way as to achieve equilibrium between the gravitational and centrifugal forces exerted upon it. The product of this interaction, known as the coriolis force, acts upon tidal streams in the northern hemisphere to deflect the flow of water to the right. Therefore, even were the Earth a smooth sphere uniformly covered with water, the tides would not flow directly towards or away from the high tide wave.

But the shape of the surface of the real Earth confines the oceans and seas to basins. The irregular shapes of the ocean basins affects the manner in which the waters can respond to the constantly-changing gravitational field. Moved by gravity but constrained by topography the water moves in a complex pattern of advance and retreat whose magnitude and timing display the phenomenon of resonance. The tides of the Severn Estuary, for instance, move up and down in very much the same way as the water in a bath will slosh back and forth when moved gently by the hand. It is surprising how little force, when it is regularly applied, is needed to create a large movement of water and it is notable that the motion settles into a steady pattern in response to the shape of the container and to the rhythm and vigour of the agitation. All tidal movements, great and small, are resonant in this way or are affected by the movements of adjacent bodies of water that are themselves in a state of resonance.

Thus it is that the tides as they respond to the forces on them, and to the influences of topography, assume a state of steady pulsation and acquire a complex and regularly repeating pattern of horizontal movement. So regular are the tides in their behaviour that the state of the tide can be predicted many years ahead with accuracy and confidence. This feature of the tides is important when considering them as a source of energy, for although they are a variable natural phenomenon the timing and extent of their variation can be foreseen accurately a long way into the future.

THE QUANTIFICATION OF TIDAL ENERGY

It is possible to tap both the horizontal and the vertical movement of tidal water as an energy source. In 1956 the French engineers Remenieras and Smaghe suggested (Bernshtein, 1961) a design for an underwater screw propellor machine to exploit the kinetic energy of water moving as a tidal stream across the sea bed. They give the power output of such a machine as:

$$P = 300\, A(V/10)^3\, \text{kW}$$

where A = swept area of rotor (m^2)
V = velocity of water (m s^{-1})

The output of such a machine would, however, be only modest because tidal streams rarely exceed 3 m s^{-1} and their speed varies sinusoidally over the period of a tide. A power output which is continuously varying is difficult to utilize. Furthermore, the engineering problems associated with a turbine submerged beneath 20 m of seawater are formidable. Consequently, the idea of using the kinetic energy of the flowing tides has not been investigated further.

Water elevated by a tide possesses by the force of the Earth's gravity potential energy since, were it to fall toward the earth, it would be capable of doing work. This form of energy, as contrasted with the kinetic energy of moving objects, is the capacity to do work that a body possesses by virtue of its position or configuration. The formula giving the potential energy of a raised body is:

$$E = mgh \text{ J}$$

where m = mass (kg)
g = acceleration due to gravity (9.8 m s^{-2})
h = height of centre of gravity (m)

The terms of this expression for an impounded body of water are illustrated in Figure 9.5. It will be observed that in the case of a tidal barrage the height h is the tidal range.

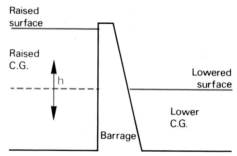

Figure 9.5 Potential Energy of an Impounded Body of Water

In Europe from at least as early as the eleventh century the potential energy of the tides has been exploited for the purpose of driving corn mills. Tidal water was impounded at the top of its cycle, after which, as the tide fell, a height difference was created which imparted potential energy to the raised body of water. Work was performed when the water was released to a lower level through a water wheel. Traditional tidal mills operated for only one period, during which the impounded water was discharged to the lowered level of the sea, in each tidal cycle. Figure 9.6 shows that this method of operation results in long periods of inactivity when for more than 50 per cent of the time the machinery is idle.

Modern tidal power plants, such as that built in 1967 on the River Rance on the north coast of

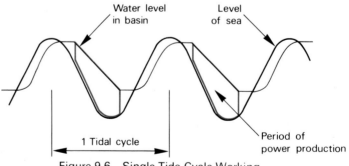

Figure 9.6 Single Tide Cycle Working

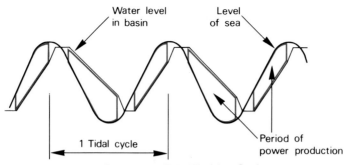

Figure 9.7 Double Tide Working Cycle

Brittany, make use of the difference in water levels produced by both the falling and the rising tide in the fashion illustrated by Figure 9.7. Power is produced when, on either the flood or the ebb, there is sufficient difference in level between basin and sea to operate the machinery. Turbines with adjustable blades are used to exploit the water as it flows in either direction.

Both single and double tide working installations extract the same amount of energy from the same tide but the output from a double tide working cycle is more continuous and therefore easier and more economical to use for power production.

During the course of the last 30 years much ingenuity has been devoted to overcoming the discontinuities and fluctuations in output that are still to be seen in Figure 9.7, and to achieving an even smoother power production curve. Proposals have been made for tidal energy schemes incorporating two, three, and sometimes four interconnecting basins. It is clear that much smoother power curves are possible, but it has also been found that all multi-basin designs pay a heavy penalty in reduced power output for their more even level of production. Multiple basin installations can produce only some 38–82 per cent of the output of the equivalent single basin project (Bernshtein, 1961). The smoother the output curve the smaller is the proportion of the energy of a site which can be captured and put to use. It can be shown (Bernshtein, 1961) that the annual gross energy production of a tidal barrage where the tidal regime is semi-diurnal is given by the following formula:

$$E = 7{\cdot}09\,Ar^2\,\text{TJ}$$
where A = area of basin (km^2)
r = average tidal range (m)

It will be noted that the size of the energy resource is proportional to the square of the average tidal range. Consequently it is necessary to position tidal energy projects on coasts where the range of tide is as large as possible. Since the energy available depends upon both the area of the basin and the range of the tide it is evident that the production of useful energy from the tide is a matter which is specific to particular geographical sites.

THE MAPS

Because the tidal range has so great an influence upon the productivity of a tidal energy site a co-range chart showing the average range of the tide can reveal those parts of the coast that are most promising for power generation. Map 9.1 shows that the greatest tidal range in the British Isles occurs in the upper reaches of the estuary of the river Severn. The Severn Estuary is, in fact, the most favoured tidal energy site in Europe. It possesses a range of tide 1·3 m greater than occurs at the barrage built across the mouth of the River Rance on the north coast of Brittany.

Average tidal ranges of 6m are found in the Solway Firth, Morecambe Bay, and in the estuaries of the rivers Mersey and Dee, while the northern end of the Menai Strait experiences a range of just over 5m. Carmarthen Bay also records a range of between 5 and 6m. Tidal ranges are smaller in Ulster, all round Scotland, and on the northeastern coast of England. The average range increases to more than 4m at the mouth of the River Humber and at the Wash a large body of water could be impounded where the range is nearly 5m. The Thames Estuary, lying so near to London, may be taken into consideration even though its area is small and the tidal range is about 4m. No tidal energy sites of promise are to be found on the south coast of England. Tidal ranges are extremely large in the Channel Islands but sites for other than very small basins are lacking. The 10 most favoured tidal energy sites in the United Kingdom are shown on Map 9.2.

Maps 9.3 to 9.12 show these 10 sites in greater detail at a uniform scale. The area shown on each map is the superficial area at mid-tide, giving the average area of impounded water. It is likely that the construction of large barrages such as these will have the effect of reducing the tidal range in their vicinity by about 10 per cent. Although the accuracy of this prediction is uncertain the tidal ranges shown on Maps 9.3 to 9.12, and also Maps 9.14 and 9.15 covering North American sites, are 10 per cent smaller than those occurring naturally. It is these reduced ranges that are used, in the following paragraphs on the size of the tidal energy resource, to calculate the energy potential of each site.

THE SIZE OF THE RESOURCE

When the 10 British tidal energy sites are evaluated by use of the formula given previously the figures in the first column of Figure 9.8 are obtained. If power plant at these sites were equipped to generate electricity with modern variable blade turbines, working on either a single basin single cycle or a single basin double cycle basis, the electricity produced would be 34 per cent of the potential energy dissipated (Bernshtein, 1961). The figures resulting are listed in the second column of Figure 9.8. The third column lists the length of barrage required to form the tidal basins. The installation on the River Rance, when assessed in the same way, is calculated to produce 1·70 PJ of electricity each year from a basin impounded behind a barrage 0·62 km long.

In 1979 the total electricity consumption in the United Kingdom was 848 PJ. If, therefore, all 10 tidal energy sites were exploited and used for electricity production, they would together provide 23 per cent of our annual requirement.

Site	Potential energy PJ	Electrical energy PJ	Barrage km
Strangford Lough	8·7	3·0	0·6
Solway Firth	102·7	36·4	29·0
Morecambe Bay	57·2	19·4	18·3
Dee and Mersey	46·7	15·9	25·2
Menai Strait	17·6	6·0	10·9
Carmarthen Bay	68·9	23·4	27·7
Severn Estuary	172·3	58·6	13·8
Thames Estuary	12·4	4·2	8·0
The Wash	56·1	19·1	19·4
River Humber	25·4	8·6	6·5
Totals	456·6	194·6	159·4

Figure 9.8 United Kingdom Tidal Energy Projects

It is instructive to rank the 10 projects in order of their electricity production per kilometre length of barrage. Although the expense and difficulty of constructing a barrage will differ with the depth of water at the site, the nature of the bottom, and a number of other factors, it is nevertheless true that the energy obtained from a length of barrage gives a good initial measure of the merit of a particular site. Figure 9.9 shows that the most meritorious tidal energy site in the United Kingdom is Strangford Lough in Northern Ireland. Only a modest annual energy production is to be expected here, because of the small range of the tide, but the barrage required is very short and construction costs would be low. Next in order of merit, and nearly 20 times as productive, is the Severn Estuary barrage. At the bottom of the list comes the Thames Estuary. A tidal energy barrage on the Thames might only be justified by the proximity of the large London energy market or by the value of a dam for flood protection purposes.

Site	Electricity output TJ per km barrage
Strangford Lough	5000
Severn Estuary	4246
River Humber	1323
Solway Firth	1255
Morecambe Bay	1060
The Wash	985
Carmarthen Bay	845
Dee and Mersey	631
Menai Strait	550
Thames Estuary	525

Figure 9.9 Project Rank Order

A criterion for these figures is provided by the turbines on the River Rance whose annual electricity production is 2742 TJ per km of barrage. It is clear that Strangford Lough, where the corresponding figure is 5000 TJ, is the site at which British tidal energy production should be initiated. An idea of the scale of construction work entailed by tidal power proposals of this type may be had from examining the longest barrage in the world, the Afsluitdijk, completed in 1962 in the Netherlands to enclose the Ijsselmeer. It is 32·5 km long, about one-fifth of the total length of barrage that would be required to develop all 10 United Kingdom tidal energy sites discussed in this chapter.

THE TIDAL ENERGY REGIME OF NORTH AMERICA

The west coast of the United States nowhere experiences tides with an average range of as much as 2 m. Further north in Canada and southern Alaska the average range just reaches 4 m. When, however, the tidal wave enters the confined waters of Cook Inlet it increases in amplitude and at Fire Island, about 15 km west of the capital city of Anchorage, its range is 7 m. A very large tidal power station could be built here capable of producing 113·3 PJ of electricity a year. By incorporating Fire Island itself into the barrage, as shown on Map 9.14, the length of artificial construction required would be 20·2 km and the project would have the high productivity factor of 5609 TJ per km of barrage.

The tidal range decreases rapidly north of the Aleutian Islands and declines to less than 1 m in the Bering Strait. Tides are slight on the north coast of Canada and in the Canadian arctic archipelago. The range remains low in Baffin Bay and Davis Strait but, as shown on Map 9.13, it increases sharply in Hudson Strait and reaches a value of 6 m at Lake Harbour on Baffin Island and in Ungava Bay on the south side of the strait. However, thick sea ice encumbers these coasts for seven months of the year and it is doubtful if tidal power plant could operate effectively under such severe conditions.

The Gulf of Mexico and the east coast of the United States is, like the west coast, an area of slight tidal activity. When, however, north and east of Cape Cod, the modest tidal wave advances into the Gulf of Maine it encounters the southern end of Nova Scotia and is divided. Part of the water flows into the Bay of Fundy and, in the enclosed upper part of the bay it rises and falls to produce the largest tides in the world. At Burntcoat Head in Minas Basin the average range is 11·25 m with a maximum spring range of no less than 14·5 m. If such large movements of water could be harnessed the Bay of Fundy could become the world's greatest tidal power house.

The bays of Cobscook and Passamaquoddy, shown on Map 9.15 at the entrance to the Bay of Fundy on its northern shore, lie either side of the border between the United States and Canada. These two bays have been the site of tidal energy study, speculation, and indeed some construction work for more than 60 years. Figure 9.10 shows that a barrage less than 7 km long built across their mouths could be used to generate a large amount of power. On the south shore of the Bay of Fundy a short barrage built to enclose Annapolis Bay would establish a small power installation of high merit. But it is at the head of the Bay of Fundy, where both tidal range and topography are favourable, that the chief tidal power resource of the North American continent is to be found.

Site	Potential energy PJ	Electrical energy PJ	Barrage km
Cook Inlet	333·3	113·3	20·2
Cobscook/Passamaquoddy	70·7	24·0	6·9
Annapolis Bay	27·5	9·4	0·9
Chignecto Bay	331·3	112·6	9·3
Minas Basin	812·9	276·4	6·4
Totals	1575·7	535·7	43·7

Figure 9.10 North American Tidal Energy Projects

Figure 9.10 shows that Minas Basin and Chignecto Bay could together generate 389 PJ of electricity a year. So powerful, in fact, are the tides entering and leaving Minas Basin that it would not be easy to accommodate the necessary turbines in such a short barrage.

Two barrages built in the positions shown on Map 9.15 could supply 3.6 per cent of the electricity consumed in 1979 by the United States and Canadian economies combined. This enormous energy resource remains untapped because the Bay of Fundy is some hundreds of kilometres from the large centres of population and industry in North America. In an age of energy scarcity, however, it is likely that they will be exploited not to feed the North American electricity grid but rather to produce hydrogen fuel by the electrolysis of water, or for the electric smelting of metal ores at or near the places where the energy is produced.

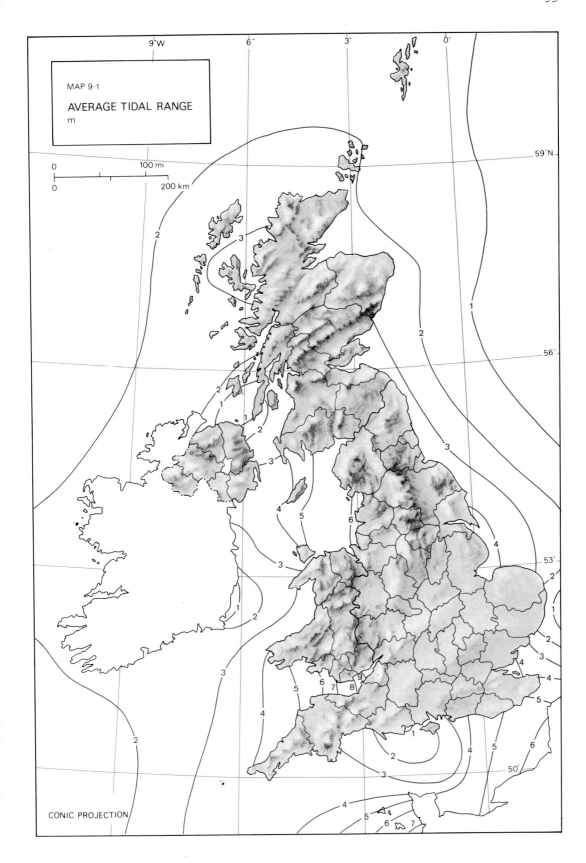

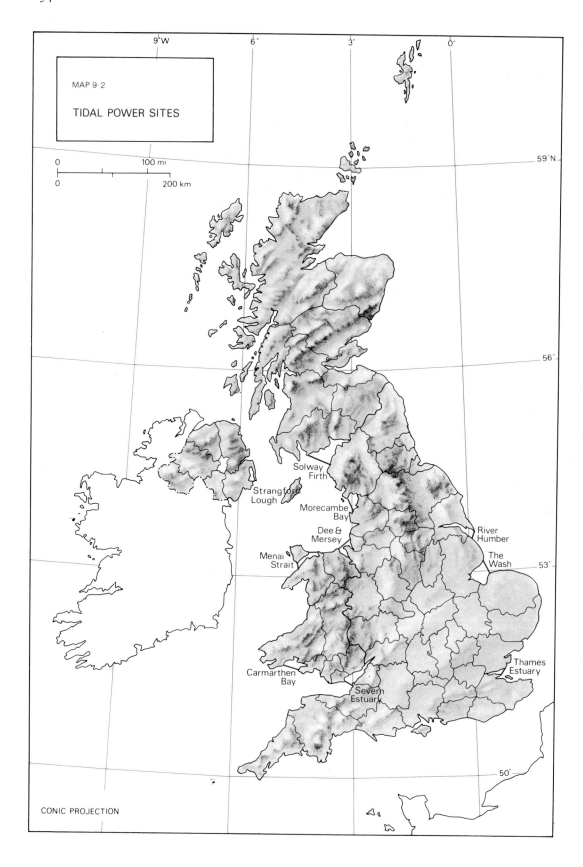

MAP 9·2

TIDAL POWER SITES

CONIC PROJECTION

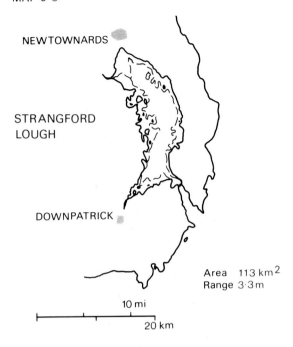

MAP 9·3

NEWTOWNARDS

STRANGFORD LOUGH

DOWNPATRICK

Area 113 km²
Range 3·3 m

10 mi
20 km

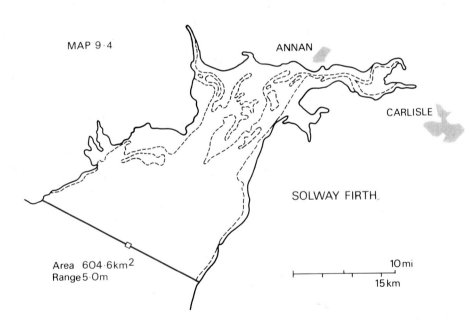

MAP 9·4

ANNAN

CARLISLE

SOLWAY FIRTH

Area 604·6 km²
Range 5·0 m

10 mi
15 km

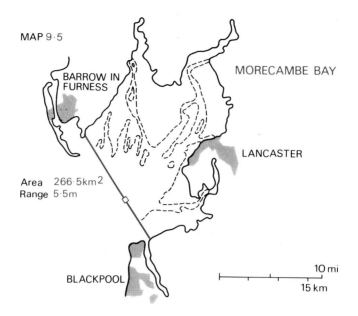

MAP 9·5

MORECAMBE BAY

BARROW IN FURNESS

LANCASTER

Area 266·5 km²
Range 5·5 m

BLACKPOOL

10 mi
15 km

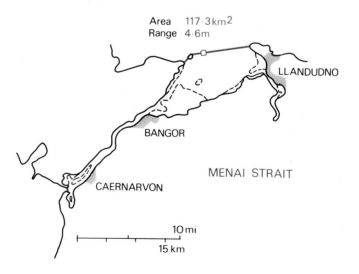

MAP 9·6

Area 117·3 km²
Range 4·6 m

LLANDUDNO

BANGOR

CAERNARVON

MENAI STRAIT

10 mi
15 km

MAP 9·7

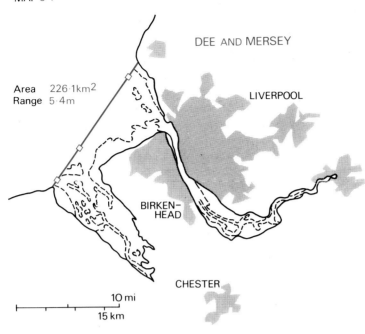

MAP 9·8

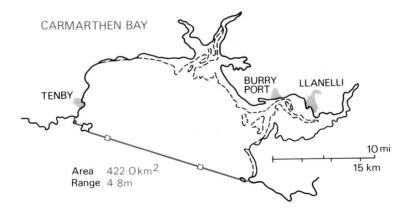

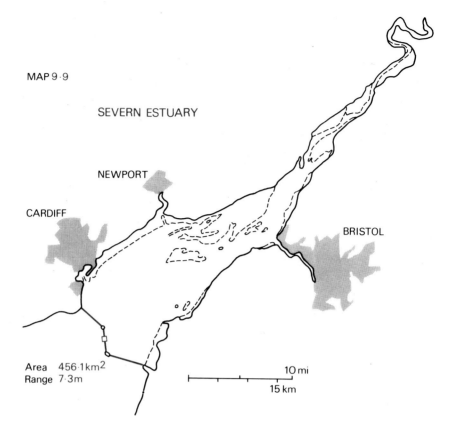

MAP 9·9

SEVERN ESTUARY

NEWPORT
CARDIFF
BRISTOL

Area 456·1 km²
Range 7·3 m

10 mi
15 km

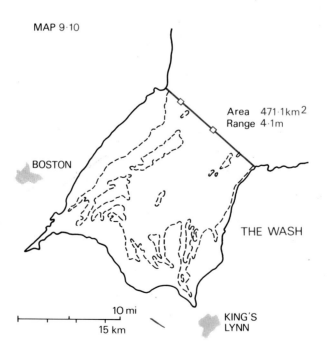

MAP 9·10

Area 471·1 km²
Range 4·1 m

BOSTON

THE WASH

KING'S LYNN

10 mi
15 km

MAP 9·11

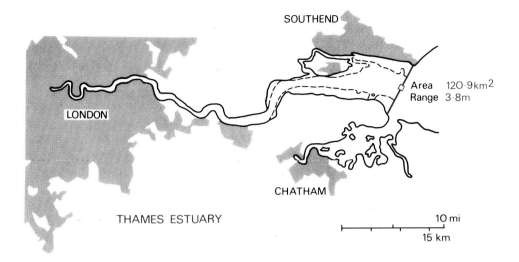

MAP 9·12

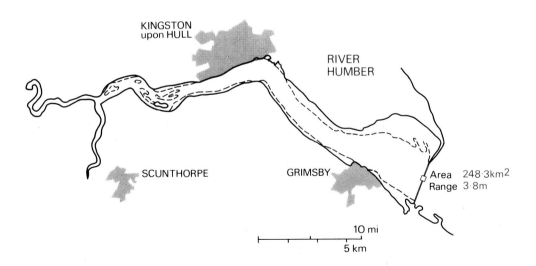

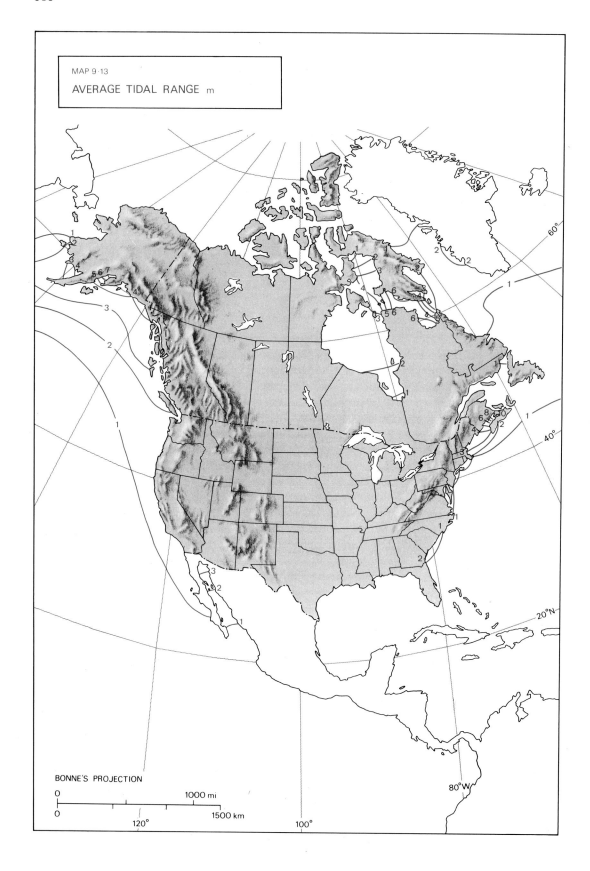

161

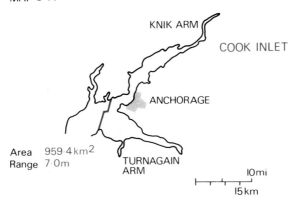

MAP 9.14

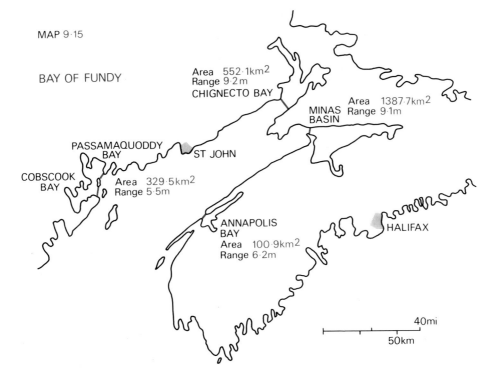

MAP 9.15

LIVERPOOL INSTITUTE OF
HIGHER EDUCATION
THE MARKLAND LIBRARY

Chapter Ten
Conclusion

It is often supposed, particularly in economic analyses, that an economy that is enjoying a rising standard of living must necessarily experience an equivalent increase in fuel consumption. It is true that in the first 80 years of the twentieth century the quantity of fuel used by the United Kingdom economy has more than doubled, from 4439 PJ in 1900 to 9209 PJ in 1979. This fact, however, conceals an important energy trend, for in the same period of time the United Kingdom gross domestic product increased from about £28,000 million to nearly £103,000 million when reckoned in constant 1975 pounds sterling. To the extent that GDP can be considered to be the criterion of our well-being, this increase shows that we are now nearly four times as well off as our forefathers at the turn of the century. Prosperity has in fact increased almost twice as fast as has fuel consumption.

The explanation for this fact is to be found in the advances that have been made in techniques of fuel use. So great have been these improvements that they have more than counterbalanced the effects of several new industries which are very extravagant of energy. Electricity generation, petrochemicals, road transport, and space heating of buildings, all rudimentary before 1900, are activities which require very large supplies of energy. But, as new ways of using fuels have arrived, so compensating improvements have occurred in the design of engines, boilers, buildings, and in all kinds of industrial plant. It is a tribute to the skills of architects, engineers, and manufacturers that fuel use has only doubled during the time when many important new industries hungry for energy have been introduced.

As a result of more effective use the amount of fuel expended to create a given quantity of goods and services is now much smaller than it was in 1900. Figure 10.1 shows the path that this trend has taken. The ordinate shows the number of megajoules of fuel that have been needed in this century in Britain to produce £1 of GDP at 1975 values. From 171 MJ in 1908 the figure has, after passing through a peak value of 173 MJ in 1924, declined to 89 MJ in 1979. The Great War, the post-war boom in the early twenties, and the General Strike of 1926 disturb the shape of the graph during the first quarter of the century, while a much smaller perturbation occurs at the time of the Second World War. But these oscillations do not disguise the general downward trend. The slope of the graph in Figure 10.1 suggests that further increases in prosperity can be expected, even in a time of energy scarcity, if the energy cost of economic activity continues to decline. It is of considerable interest to know how far this line of development can be taken.

In an attempt to answer this question the United Kingdom economy has been subjected to a detailed and thorough analysis by a team working under the auspices of the International Institute for the Environment and Development. Their report (Leach *et al.*, 1979) divides the economic life of the nation into nearly 400 categories grouped under the principal headings of industry, the domestic, commercial, and institutional sectors, and transport. Energy conservation measures applicable to each category are assessed in order to discover the fuel needs, and not merely the projected fuel demand, of economic life in a future energy-conscious society. Only well known and properly understood methods of energy conservation are taken into account. Basic oxygen steel making is more economical of energy than are open hearth furnaces, for instance, while combined heat and power plants make much better use of fuel than do existing electricity-only power stations. Better designed buildings can at the same time improve comfort and reduce

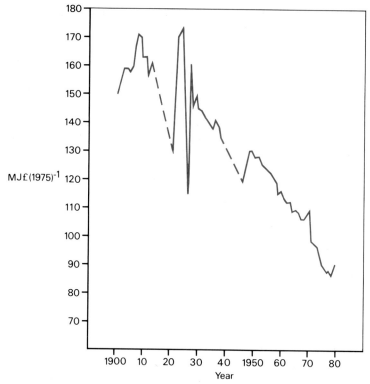

Figure 10.1 Energy Cost of Economic Activity in the United Kingdom

energy consumption while, in the transport sector, improvements in the design of cars, trains, and aircraft can bring about large reductions in fuel use. The IIED study assumes that only a small contribution to future energy needs will come from renewable energy resources.

Energy conservation measures of the kind considered in the IIED report cannot be brought into effect quickly. Industrial plant, roads, railways, bridges, and vehicles wear out and need to be replaced only in due time. The factor which exerts the most severe limit on the speed with which change can proceed is the long period of time required to renew the nation's stock of buildings. The useful life of a building lasts, on average, about 60 years. Energy supply systems, also, can be adapted only slowly. The IIED report therefore looks to the future and examines the economic life and fuel needs of the United Kingdom between the present time and the year 2025.

The effects of two trends in economic activity are investigated. GDP in 2025 is assumed in the high growth case to be three times its present size and in the low case the level of economic activity is assumed to have only doubled. The economic future in Britain is shown in the report to be one of increasing material prosperity. The report makes an assessment of the fuel needs of the year 2025 by summing the expected requirements of the 400 consumer categories, for both the high and the low growth cases, to arrive at a total national fuel demand. The result is unexpected. Energy demand in the low case at 7935 PJ is less than present consumption, while in the high case a tripling of GDP by 2025 could be achieved using no more fuel than was consumed in 1979. In effect, the energy cost of economic activity would decline to roughly 40 MJ £$(1975)^{-1}$ in the low case while in the high case only about 30 MJ would be needed for £1 of wealth produced. It is clear that energy conservation, applied on a national scale, can produce remarkable benefits for our future economic life.

The world's remaining stocks of fossil fuels can be effectively husbanded by thoroughgoing conservation measures but, unless civilization collapses, a need for large supplies of primary energy will remain. How is this reduced, but still indispensable, supply to be obtained? The preceding chapters have shown that ocean thermal energy conversion is impossible in the seas surrounding the British Isles. A contribution to supplies from the geothermal resource can be hoped for but, in the present state of knowledge, its size cannot be established. Solar energy in the British climate is unlikely to be capable of being converted into electricity and this resource therefore cannot be expected to contribute to the flow of fuel through the economy. Solar energy has, however, a large role to play in the future economy of the United Kingdom. It is well adapted to act as a fuel saver by furnishing heat directly for the provision of hot water and space heating of buildings.

The five remaining renewable resources, as has been shown in the preceding chapters, are capable of the following levels of energy production:

Wind	555 PJ	electricity
Waves	600 PJ	electricity
Rivers	24 PJ	electricity
Biofuels	227 PJ	gas
Tides	195 PJ	electricity

It is fortunate that all but one of these resources, biofuels, are best harnessed for electricity production. The techniques at present used for electricity generation at thermal power stations are the most wasteful of all fuel-consuming industries. Electricity from renewable resources can therefore be introduced into the economy at the point where improvement in energy performance is most needed. They can, by substituting for thermal generation, bring about very large reductions in fuel use and also reduce the amount of waste heat rejected to the environment.

In Figure 10.2 the supply potential of renewable energy resources in the United Kingdom has been set against the level of energy demand predicted for the year 2025 by the low growth case of the IIED report. The flow of energy through the economy under these circumstances is illustrated. The diagram follows the same graphical conventions as are used in the description of the 1979 United Kingdom economy given in Figure 1.3 in the introductory chapter.

Certain features of this diagram call for comment. In the first place, a power system utilizing renewable energies must acknowledge the fact that the forces of nature do not run according to a regular daily timetable. River energy, a small resource in Britain, is the only renewable resource that can, by means of barrages, be regulated precisely to suit any pattern of demand. Around the sea coasts tidal energy flows in accordance with a regular cycle which is accurately predictable even years ahead, but whose timing is out of phase with normal social routines. Wind, and therefore waves, will deliver electricity in a manner whose timing cannot be predicted reliably. If these sources of energy are to be tapped effectively the economy must be able to accommodate a mismatch between the timing of demand and the availability of supplies. We have been accustomed for about a century now to obtaining power exactly when we want it, and the prospect of a poor fit between availability and demand is therefore disturbing. There is little cause for alarm, however. The economy depicted in Figure 10.2 contains three features calculated to bring into relation a regular pattern of demand and a system of supply relying partly upon renewable resources.

Nearly half the electricity generated in the year 2025 would be used to produce hydrogen by the electrolytic dissociation of water. Hydrogen gas can be delivered to consumers as a fuel in the normal way through a piped distribution network. Energy losses are incurred at hydrogen plants because the production process is only about 60 per cent energy efficient (McAuliffe, 1980). These losses, and some technical difficulties involved in the handling of a low molecular

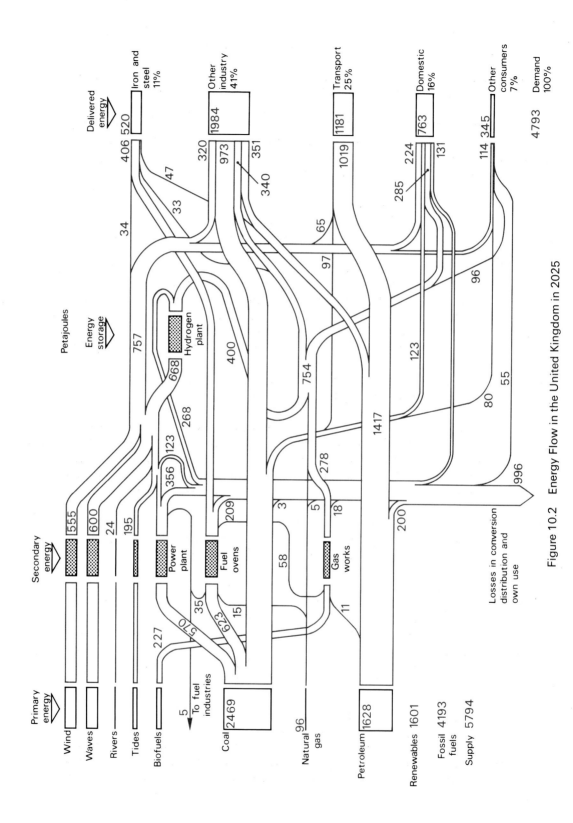

Figure 10.2 Energy Flow in the United Kingdom in 2025

weight gas like hydrogen, are inescapable if the energy produced by wayward natural forces is to be stored as a fuel for use later at convenient times. Almost exactly half the 2025 demand for gas can be met by manufactured hydrogen.

A biofuels industry, making use of all available organic material together with waste from an enlarged forest estate, could by the year 2025 supply 227 PJ of energy a year. Biogas and manufactured hydrogen could jointly furnish nearly all the gas required by a future energy-conscious British economy. These two new gas industries will be needed, for the United Kingdom's reserves of natural gas, which in 1979 supplied nearly a fifth of all primary energy, are expected to be almost exhausted by the end of the first quarter of the twenty-first century.

The IIED report assumes that by 2025 electric road vehicles will have penetrated the parts of the market for which electric propulsion is suitable. In the low growth case it is assumed that 30 per cent of car travel and 40 per cent of van and bus mileages are electric powered. The batteries of this fleet of electric vehicles form a large store of electrical energy and they can perform part of the function of smoothing supply to demand as well as economizing primary energy.

The third energy smoothing feature of the economy described in Figure 10.2 is the small part of the electricity supply which would continue to be produced at coal-fired power stations. Their output is, of course, capable of close control to follow fluctuations in the timing and level of electricity demand. The Britain of 2025 could thus be provided with three ways of accepting and using effectively the variable energy supply that can be obtained from renewable resources.

About one-third of the flow of waste heat rejected to the environment by the United Kingdom economy in 2025 would originate in thermal power stations, with nearly another third coming from hydrogen plants. Combined heat and power (or CHP) plants are assumed to be making some contribution to power supplies by that time, and therefore all the heat rejected by power stations would not be lost. Heat distributed from CHP plants and other sources is shown as providing 17 per cent of total energy demand in the domestic sector and 16 per cent of that needed by other consumers. After extracting these flows of 131 and 55 PJ respectively for use, the waste energy rejected by the economy in 2025 amounts to 996 PJ. The corresponding figure for the 1979 economy, upon which conservation measures and supplies from renewable resources have made but little impact, is 2764 PJ. It is sobering to reflect upon the immense waste revealed by the difference between these two figures.

On the continent of Europe, and particularly in Scandinavia, it is common practice to build power stations in such a way as to provide the consumer with electricity and also with heat for industrial processes and space heating of buildings. The efficiency of electricity generation in these CHP installations is only some 24 per cent because the temperature of the exhaust steam is about 85 °C rather than the 38 °C usual in electricity-only power plants. But the use found for the heat rejected by the turbines raises the thermal efficiency of the whole system to 70 per cent. This more than doubles the performance at present achieved in Britain by electricity-only plants, and it is nearly four times the average efficiency of power station operation in North America. It is clear that were European practice to be emulated in the United Kingdom and North America large reductions in fuel consumption could be made. What are the principal obstacles that have hindered the introduction of CHP in Britain?

A search for an explanation leads the enquirer to one of the early paragraphs of the 1957 Electricity Act. This Act of Parliament, the most recent in a series of similar statutes, imposes upon the Central Electricity Generating Board the duty to 'develop and maintain an efficient, co-ordinated and economical system of supply of electricity in bulk to all parts of England and Wales'. The 1947 Electricity Act places identical duties upon the regional generating boards who are responsible for producing electricity in their areas. Similar acts govern electricity supply in Scotland and Ulster, and none mention a duty to supply energy in a form other than electricity. These narrow terms of reference discourage the electricity industry from building and operating CHP plants. To do so would result in a lower efficiency in the electrical part of a combined energy supply system, and so could be said to be contrary to the intention of the 1947 and 1957 Acts.

The increased need to use fuel efficiently in the generation of electricity has been recognized in recently published official proposals for reorganizing the industry (Department of Energy, 1978). The electricity industry's terms of reference would be widened to include 'the development of methods by which heat obtained from or in connection with the generation of electricity may be used for the heating of buildings or for any other useful purpose'. If implemented measures of this kind would have far-reaching effects upon the nation's fuel requirements. However, the proposal also contains a lame qualification to the effect that a new Electricity Corporation need only 'have regard to such requirements of national energy policy as may from time to time be communicated to it by the Secretary of State'. A less lackadaisical reform of the legislation is possible and needed. Parliament has the power, while apparently lacking the will, to make a striking improvement in the energy efficiency of the British economy by modernizing the Electricity Acts and broadening their scope to include the duty to supply heat as well as electricity to the consumer.

The description of the future that emerges from Figure 10.2 gives good ground for optimism. The fuel requirement of 4193 PJ in the year 2025 is slightly less than half the amount consumed in the United Kingdom in 1979. Nevertheless, energy delivered to consumers in 2025, totalling 4793 PJ, provides enough power for an economy twice the present size. An energy-conscious future, therefore, can in Britain be a prosperous one.

The economies of both Canada (Energy Policy Sector, 1976) and the United States (CONAES, 1979) have, like the United Kingdom economy, been studied in order to discover the scope of practical energy conservation measures and to try to chart a course towards an energy-conscious future. The North American studies, like the IIED report on the future of the British economy, foresee steadily rising levels of material prosperity despite only a slight increase in the use of fuel. The EPS report assumes that the Canadian economy will have doubled in size by the year 1990 while one of the scenarios in the CONAES report, scenario B, supposes that economic activity in the United States will increase twofold by 2010. The statistics on distributed energy shown on Figure 10.3 are assembled from these two studies, but the assumption is made that economic growth will be somewhat slower than the reports expect, and that the doubling of the North American GDP will be postponed until the year 2025.

Primary energy supplies shown in Figure 10.3 include a contribution of 17 EJ from renewable resources. This is nearly twice the entire quantity of energy which entered the United Kingdom economy in 1979, but even so large a supply of renewable energy would only account for about 18 per cent of North American requirements in the year 2025. The biofuels resource could help to reduce the consumption of natural gas to less than one quarter of the 1979 level but very large quantities of coal would still be needed for electricity generation and for industry. The electricity generating renewables are estimated to be able to make little contribution to the transport sector, for which a huge demand for oil is assumed to continue into the second quarter of the twenty-first century. The waste from this relatively energy-conscious economy is, by coincidence, almost exactly the same as the amount of energy it would gain from renewable sources of supply. Waste energy in 2025 is, however, less than half that rejected to the North American environment by the economy of 1979.

A full exploitation of the renewable energy resources of North America would reduce the consumption of fossil fuels to 77 per cent of the amount used in 1979. By comparison, in the United Kingdom the demand for fuel in 2025 could be less than half the 1979 figure. Both these reckonings leave out of account the contribution that could be obtained from solar energy, and it is likely solar has an important part to play in the sunnier climate of North America. Nevertheless, it is surprising to discover that the United Kingdom is, in proportion to need, better endowed with renewable energy resources than is the continent of North America.

There are no insuperable geographical or technical obstacles, either in Britain or North America, to harnessing renewable energy resources on a very large scale. An economy powered in this way would be robust, safe, flexible, and independent. But renewable energies are, in official opinion, everywhere dismissed as of only marginal significance. Research into renewables, it is true, receives limited governmental support in many countries but little or no action follows from the new knowledge and skills that emerge. Two misconceptions are responsible for this neglect and inaction.

In most industrialized countries official opinion assumes that future energy supply problems will be solved by creating a nuclear power industry. In Britain, for example, a garbled version of

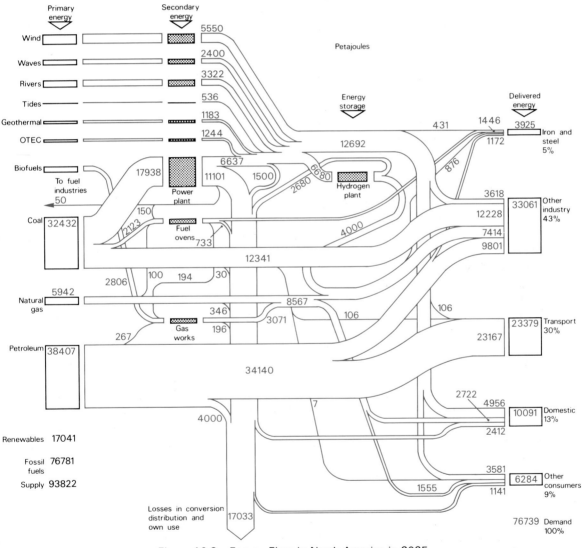

Figure 10.3   Energy Flow in North America in 2025

the argument that nuclear power is inevitable has appeared recently (Secretary of State for Energy, 1978). This and other similar documents ignore, however, the fatal flaw in the argument for nuclear power. This is that even the highest possible standards of engineering competence and conduct are too fallible to manage such a dangerous technology. Perfection is required, but cannot be obtained.

Public opinion, at the time of writing, is acquiescent about the nuclear power industry and a certain apprehensive optimism characterizes the common view of a nuclear-powered future. But the first nuclear catastrophes will alter popular opinion and the politics of energy will change in consequence. Although accidents are not the only threat posed by nuclear power, it is likely that a serious incident will be the event responsible for foreclosing the nuclear option. To abandon nuclear power would, however, carry with it no economic penalty. Figures 1.2 and 1.3 show that the contribution of nuclear power to the British and North American economies in 1979 was small. Its removal will cause only a slight fall in energy supplies and a small and transient perturbation in economic life. The future economies depicted in Figures 10.2 and 10.3 do not make use of nuclear power.

The second cause of official scepticism about the usefulness of renewable energies is the conviction that they will always be too expensive to exploit (ACORD, 1982). It is without doubt true that fossil fuels, at their present prices, furnish energy more cheaply than can biofuels, solar energy, or the electricity generating renewables. In this sense they are indeed too expensive to use. But present price relationships will not continue into the future unchanged. As fossil fuels become scarcer and more difficult to obtain their price will rise more quickly than the general rate of inflation. Figure 10.4, showing the price movement of industrial fuels during the 25 years to 1980, illustrates the beginning of this process in Britain. Prices in Figure 10.4 are given at constant 1975 values in order to allow for inflation. Gas, which in 1956 was manufactured from oil and coal, showed a steep decline in real price until 1974. This has followed from the exploitation of new natural gas fields in the North Sea and the consequent phasing out of manufactured gas as a fuel. During the 1950s fuel oil also declined in price although at a much more modest rate than gas.

The price of fuel oil began to rise in 1973 following the establishment of a cartel by the Organisation of Petroleum Exporting Countries, but Figure 10.4 shows that it was some 10 years before that oil ceased to become cheaper in real terms. Oil was the same price in 1973 as 1963 because, worldwide, oilfield discoveries were not keeping pace with increases in consumption during that decade. Coal, which is a more difficult fuel to use than oil or gas, has exhibited less price volatility over the last 25 years than the other two fuels. Coal prices tend to follow and reflect the price trend of oil and gas.

Since 1973 the real cost of coal and oil has risen to a higher level than 1956 while natural gas prices are also rising steeply. Some changes in the rate of increase are to be expected, but it is clear that the recent trend shown in Figure 10.4 cannot continue indefinitely unless a breakdown in industrial society is to be foreseen. Unrestrained rises in fuel costs would ultimately be ruinous to industry by depriving it of its motive power. It is of great importance, therefore, to discover if there is any factor acting to place a limit on the rise in price of fuel. In fact, a limit will be reached when the cost of energy obtained from fossil fuels equals or begins to exceed the cost of utilizing renewable energy sources.

Figure 10.5 shows the trend taken by the price of electricity in the United Kingdom during the 30 years to 1980. These prices, like those used in Figure 10.4, have been discounted to 1975 values. Electricity, like fossil fuels, became steadily cheaper in real terms until the early 1970s, but since that time it too has become more expensive. Also shown on Figure 10.5 are three recent estimates of the unit cost of generating electricity from the tides at a barrage built across the Severn Estuary. These three points and the curve of the graph cannot, of course, be compared

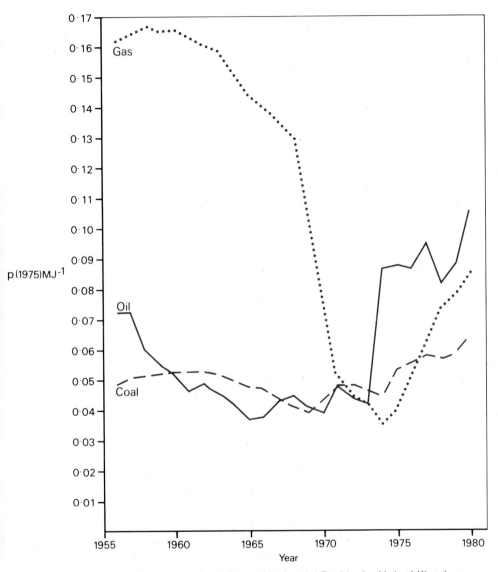

Figure 10.4 Real Price of Industrial Fuel in the United Kingdom

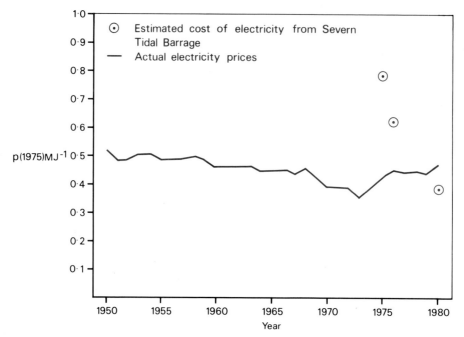

Figure 10.5   Real Price of Electricity in the United Kingdom

directly because generating costs and sales prices are not the same thing. However, the evidence shows that electricity prices are increasing at a time when tidal electricity costs are declining. At present it cannot be foreseen when the point of coincidence of these two trends will be reached, but it seems clear that in due course it is thermal electricity rather than the electricity producing renewables that will be recognized as being uneconomic. Acting in this way, economic forces will place a limit upon the price of fossil fuels.

Figure 10.4 suggests that first oil and then natural gas prices will rise to match the cost of the renewables. For this reason thermal power plant operating in the year 2025 are shown in Figures 10.2 and 10.3 as fuelled by coal only. As that time approaches it becomes increasingly important to assess the potential of renewable energy resources as fully and accurately as possible. This atlas is offered as a contribution to our stock of knowledge about their distribution and availability in the United Kingdom and in North America.

Appendix 1
Conversion Factors

	To convert	To	Multiply by	Reciprocal
Length	inch	metre	$2 \cdot 540 \times 10^{-2}$	$3 \cdot 937 \times 10^{1}$
	foot	metre	$3 \cdot 048 \times 10^{-1}$	$3 \cdot 281$
	yard	metre	$9 \cdot 144 \times 10^{-1}$	$1 \cdot 094$
	statute mile	metre	$1 \cdot 609 \times 10^{3}$	$6 \cdot 215 \times 10^{-4}$
	nautical mile	metre	$1 \cdot 852 \times 10^{3}$	$5 \cdot 400 \times 10^{-4}$
Area	$inch^2$	$metre^2$	$6 \cdot 452 \times 10^{-4}$	$1 \cdot 550 \times 10^{3}$
	$foot^2$	$metre^2$	$9 \cdot 290 \times 10^{-2}$	$1 \cdot 076 \times 10$
	$yard^2$	$metre^2$	$8 \cdot 361 \times 10^{-1}$	$1 \cdot 196$
	acre	$metre^2$	$4 \cdot 047 \times 10^{3}$	$2 \cdot 471 \times 10^{-4}$
	statute $mile^2$	$metre^2$	$2 \cdot 590 \times 10^{6}$	$3 \cdot 861 \times 10^{-7}$
	nautical $mile^2$	$metre^2$	$3 \cdot 430 \times 10^{6}$	$2 \cdot 915 \times 10^{-7}$
Volume	$inch^3$	$metre^3$	$1 \cdot 639 \times 10^{-5}$	$6 \cdot 101 \times 10^{4}$
	$foot^3$	$metre^3$	$2 \cdot 832 \times 10^{-2}$	$3 \cdot 531 \times 10$
	$yard^3$	$metre^3$	$7 \cdot 645 \times 10^{-1}$	$1 \cdot 308$
	gallon/(Imperial)	$metre^3$	$4 \cdot 546 \times 10^{-3}$	$2 \cdot 200 \times 10^{2}$
	gallon/(U.S.)	$metre^3$	$3 \cdot 785 \times 10^{-3}$	$2 \cdot 642 \times 10^{2}$
Mass	pound	kilogram	$4 \cdot 536 \times 10^{-1}$	$2 \cdot 205$
	ton (long)	kilogram	$1 \cdot 016 \times 10^{3}$	$9 \cdot 843 \times 10^{-4}$
	ton (short)	kilogram	$9 \cdot 072 \times 10^{2}$	$1 \cdot 102 \times 10^{-3}$
Time	minute	second	$6 \cdot 000 \times 10$	$1 \cdot 667 \times 10^{-2}$
	hour	second	$3 \cdot 600 \times 10^{3}$	$2 \cdot 778 \times 10^{-4}$
	day	second	$8 \cdot 640 \times 10^{4}$	$1 \cdot 157 \times 10^{-5}$
	month (30 d)	second	$2 \cdot 592 \times 10^{6}$	$3 \cdot 858 \times 10^{-7}$
	month (31 d)	second	$2 \cdot 678 \times 10^{6}$	$3 \cdot 734 \times 10^{-7}$
	year (365 d)	second	$3 \cdot 154 \times 10^{7}$	$3 \cdot 171 \times 10^{-8}$
Velocity	feet per second	$metre/second^{-1}$	$3 \cdot 048 \times 10^{-1}$	$3 \cdot 280$
	miles per hour	$metre/second^{-1}$	$4 \cdot 470 \times 10^{-1}$	$2 \cdot 237$
	knot	$metre/second^{-1}$	$5 \cdot 144 \times 10^{-1}$	$1 \cdot 944$
Temperature	fahrenheit	celsius	$(F-32)/1 \cdot 8$	$(C \times 1 \cdot 8) + 32$
	kelvin	celsius	subtr. $273 \cdot 15$	add $273 \cdot 15$
Energy	erg	MJ	10^{-13}	10^{13}
	calorie	MJ	$4 \cdot 187 \times 10^{-6}$	$2 \cdot 388 \times 10^{5}$
	BTU	MJ	$1 \cdot 055 \times 10^{-3}$	$9 \cdot 479 \times 10^{2}$
	therm	MJ	$1 \cdot 055 \times 10^{2}$	$9 \cdot 479 \times 10^{-3}$
	quad	MJ	$1 \cdot 055 \times 10^{12}$	$9 \cdot 479 \times 10^{-13}$
	kWh	MJ	$3 \cdot 600$	$2 \cdot 778 \times 10^{-1}$

	To convert	To	Multiply by	Reciprocal
	mtce	MJ	$2 \cdot 690 \times 10^{10}$	$3 \cdot 717 \times 10^{-11}$
	mtoe	MJ	$4 \cdot 480 \times 10^{10}$	$2 \cdot 232 \times 10^{-11}$
Energy density	BTU ft^{-2}	MJ m^{-2}	$1 \cdot 135 \times 10^{-2}$	$8 \cdot 811 \times 10$
	Langley	MJ m^{-2}	$4 \cdot 185 \times 10^{-2}$	$2 \cdot 389 \times 10$
Power density	heat flow units	mW m^{-2}	$4 \cdot 185 \times 10$	$2 \cdot 389 \times 10^{-2}$

Appendix 2
Net Thermal Value of Fuels

Fuel	Unit	United Kingdom GJ	North America GJ
Hard coal	tonne	27·79	26·67
Brown coal	tonne	—	14·27
Lignite	tonne	—	13·73
Patent fuel	tonne	26·62	29·30
Coke	tonne	26·62	28·05
Residual fuel oil	tonne	41·23	41·99
Petrol	tonne	44·58	44·79
Jet fuel	tonne	44·08	45·42
Diesel fuel	tonne	43·28	43·62
Paraffin	tonne	44·20	43·95
Methanol	tonne	22·30	22·30
Fuel wood	tonne	15·00	15·00
Natural gas	$m^3 \times 10^3$	35·15	34·73
Biogas,			
by oxygen gasification	$m^3 \times 10^3$	12·00	12·00
by anaerobic digestion	$m^3 \times 10^3$	27·00	27·00

Source: Ader *et al.* (1978)
        Ader and Wheatley (1979)
        International Energy Agency (1977)
        International Energy Agency (1981b)

Appendix 3
Multiples

10^{18}	exa	E
10^{15}	peta	P
10^{12}	tera	T
10^{9}	giga	G
10^{6}	mega	M
10^{3}	kilo	k
10^{2}	hecto	h
10	deca	da
10^{-1}	deci	d
10^{-2}	centi	c
10^{-3}	milli	m
10^{-6}	micro	μ
10^{-9}	nano	n
10^{-12}	pico	p
10^{-15}	femto	f
10^{-18}	atto	a

Appendix 4
Derivation of Maps

All the maps appearing in this atlas have been drawn for the purpose. They are based upon information obtained from the following sources.

Map 2.1	Collingbourne (1975), lists I and II.
2.2–2.15	Cowley (1978) extended to Shetland, Orkney, Ulster, and Channel Islands from Collingbourne (1975).
2.16	Map 2.15 by multiplication.
2.17 and 2.18	Cowley (1979) Figures 7 and 8, extended to Shetland, Orkney, and Channel Islands by author's method of calculation.
2.19	Baldwin (1968) page 70, and Titus and Truhlar (1969) Chart 13.
Map 3.1	Lacy (1977) Figure 31.
3.2	Allen and Bird (1977) Figure 1.
3.3	Lacy (1977) Figure 86.
3.4	Shellard (1968).
3.5	Baldwin (1968) page 74 and Canadian Normals (1975).
Map 4.1	Winter (1980).
4.2	Derived from Map 4.1
4.3	Derived from Map 4.2
4.4	Chelton *et al.* (1981).
4.5	Derived from Map 4.4
Map 5.1	Hydrographic Department (1976).
5.2	Isopleths in the Pacific Ocean from NORPAC (1960) Figure 37 and Robinson (1976) Figure 145. Isopleths in the Atlantic Ocean from Oceanographic Office (1976) Figures 11–71 and Robinson *et al.* (1979) Figure 169.
Map 6.1	Derived from Figure 6.2.
6.2	Derived from Figure 6.5.
Map 7.1	Figure 7.3.
7.2–7.4	Figure 7.5.
7.5	Figure 7.7.
7.6	Figure 7.9.
7.7–7.10	Figure 7.11.
7.11	Figure 7.14.
7.12–7.15	Figure 7.16.
7.16	Figure 7.17.
Map 8.1	Dunning (1966) and Taylor *et al.* (1971) Figure 2.
8.2	Haenel (1980) Plate 2.
8.3	King (1969).

8.4	Garland and Lennox (1962), Hyndman *et al.* (1979), Jessop and Lewis (1978), Judge (1974), Lewis *et al.* (1979) and Sass *et al.* (1981) Figures 13.16 and 13.18.
Map 9.1	Hydrographic Department (1971), (1974a), (1974b), and (1979a).
9.2–9.12	Ordnance Survey (1982).
9.13	Naval Weather Service Detachment (1974) and (1977), Hydrographic Department (1979b) and (1979c), and Godin (1980).
9.14–9.15	Bernshtein (1961) Figures 9.1 and 9.9.

Appendix 5
Sources for Figures and Diagrams

All the figures and diagrams appearing in this atlas have been compiled and drawn for the purpose. They are based upon information obtained from the following sources.

Figure	1.1	Parker *et al.* (1982) and Slessor and Lewis (1979).
	1.2	Lindal (1973) and Parker *et al.* (1982).
	1.3	Department of Energy (1980).
	1.4	International Energy Agency (1981a).
	1.6	Adapted from Saunders (1981). Volume of Repository from Roberts (1979).
Figure	2.1	Solar spectrum from Thekaekara and Drummond (1971).
	2.2	Ground level curves from Hatfield *et al.* (1981).
	2.4	Hatfield *et al.* (1981).
	2.5	Meinel and Meinel (1977).
	2.6	Long (1976).
	2.8	Transmittance of glass from Duffie and Beckman (1974).
	2.10	Compiled from DoE (1980), Bakke *et al.* (1975), and Bush and Chadwick (1979).
Figure	3.1	Calculated by method given in Allen and Bird (1977) supplemented by data from Lacy (1977).
	3.2	Averages of Allen and Bird (1977) Figure 4.
	3.3	Tagg (1957).
	3.6–3.10	Shellard (1968) and Tagg (1957).
Figure	4.2	Bowditch (1966).
	4.3	Dawson (1979).
	4.4	Bigelow and Edmondson (1947).
	4.5	Pierson *et al.* (1955).
	4.6	Converted from Salter (1974).
Figure	5.1	Diffuse curve from Burt (1954), direct curve from Jerlov (1976).
	5.2	Sverdrup *et al.* (1942).
	5.3	Curves A and B from Deitrich (1950), remainder from Marcus *et al.* (1973).
	5.4	Locations A and B from Deitrich (1950), locations C and D from Oceanographic Office (1976).
	5.5	Marcus *et al.* (1973) and Sverdrup *et al.* (1942).
	5.6	Consumption statistics from Edison Institute (1980).
Figure	6.1	Sutcliffe *et al.* (1975) and Water Data Unit (1978).
	6.2	Lawrie *et al.* (1944) and Wilson *et al.* (1980).
	6.3	Consumption statistics from Central Statistical Office (1981a).
	6.4	CNCIHD (1972) and Iseri and Langbein (1974).

	6.5	Maxson et al. (1972) and Tung (1983).
	6.6	Consumption statistics from International Energy Agency (1981a).
Figure	7.1	Cooper (1975) and Hughes (1977).
	7.2	Best (1981).
	7.3	Areas from Centre for Agricultural Strategy (1980) and Central Statistical Office (1981a).
	7.4	Arisings from Centre for Agricultural Strategy (1980), volatile solids and gas productivity from Ader et al. (1978) and Paltz and Chartier (1980).
	7.5, 7.7, 7.9, and 7.11	Areas from Department of Agriculture and Fisheries for Scotland (1980), Ministry of Agriculture, Fisheries and Food (1981a), Ministry of Agriculture, Fisheries and Food (1981b), and Welsh Office (1980).
	7.6	Arisings from Paltz and Chartier (1980) and Rutherford (1976), volatile solids and gas productivity from Ader et al. (1978) and Paltz and Chartier (1980).
	7.8	Arisings from Francis (1980), volatile solids and gas productivity from Ader and Wheatley (1979).
	7.10 and 7.12	Arisings from Paltz and Chartier (1980), volatile solids and gas productivity from Ader and Wheatley (1979).
	7.14	Gas statistics from Central Statistical Office (1981a).
	7.15	Bureau of the Census (1981) and Munn et al. (1980).
	7.16	Areas from Bowen (1978), Forest Service (1977), United States Department of Agriculture (1981), Statistics Canada (1979a), and Statistics Canada (1980).
	7.17	Mains gas statistics from Schlesinger et al. (1980) and Statistics Canada (1979b).
Figure	8.1	Brown and Musset (1981).
	8.2	Geothermal Energy Project (1982).
	8.3	Oxburgh et al. (1977).
Figure	9.2	Doodson and Warburg (1941).
	9.4	Doodson and Warburg (1941).
Figure	10.1	Cost of energy from Alford et al. (1971), Central Statistical Office (1981b), and Department of Energy (1981). Discounting from Alford et al. (1971) and Central Statistical Office (1981b).
	10.2	Delivered energy statistics from Leach et al. (1979).
	10.3	Delivered energy statistics from CONAES (1979) and Energy Policy Sector (1976).
	10.4	Cost of energy from Department of Energy (1973) and Department of Energy (1981). Discounting from Central Statistical Office (1981b).
	10.5	Electricity prices from Department of Energy (1973) and Department of Energy (1981). Tidal energy prices from Select Committee on Science and Technology (1977a), Select Committee on Science and Technology (1977b), and Severn Barrage Committee (1981). Discounting from Central Statistical Office (1981b).

Appendix 6
United Kingdom Counties

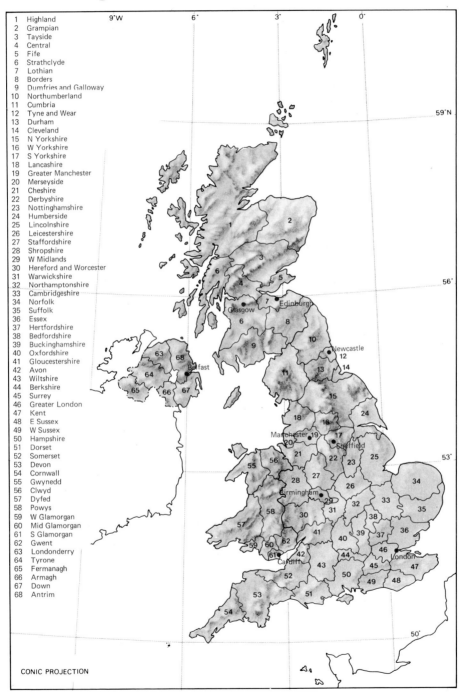

Appendix 7
North American States and Provinces

Appendix 8
Copyright Acknowledgements

All the maps and illustrations appearing in this atlas have been specially drawn for the purpose. Many incorporate, and some are entirely composed of, information which is copyright and I am indebted to the following copyright holders for permission to use their material. In some cases the copyright material makes up only a part of the illustration referred to.

The British Nuclear Energy Society for Figure 1.6.
Macmillan Journals Ltd for Figures 2.1 and 4.6, and for Maps 4.1 and 4.4.
The International Solar Energy Society for Figures 2.2 and 2.4, and for Maps 2.17 and 2.18.
Addison-Wesley Publishing Company for Figure 2.5, from Meinel and Meinel, *Applied Solar Energy*, © 1977. Addison-Wesley, Reading, MA. Figures 2.18, 2.19, 2.20 and 2.22. Reprinted with permission.
The Controller of Her Majesty's Stationery Office for Figures 2.6, 3.2 and 8.2, for Maps 3.1, 3.2 and 3.3, and for the text within parentheses in Chapter 10. Reproduced by permission of the Controller, HMSO. Crown copyright.
Environment Canada Atmospheric Environment Service for Map 2.19.
ERA Technology Ltd for Figures 3.3 and 3.6.
United States Defense Mapping Agency Topographic Centre for Figure 4.2
Ministry of Defence Hydrographic Department for Figures 9.2 and 9.4, Crown copyright reserved.
United States Naval Oceanographic Office for Figures 4.4 and 4.5, and for Map 5.2.
Prentice-Hall Inc. for Figures 5.2 and 5.5K, from Sverdrup, Johnson and Fleming, *The Oceans: Their Physics, Chemistry, and General Biology*, © 1942, renewed 1970, pp. 105, 131. Reprinted by permission of Prentice-Hall, Inc., Englewood Cliffs, N.J.
University of California Press for Map 5.2.
The Director of the Institute of Geological Sciences for Map 8.1, Crown copyright reserved.
The Director of the Ordnance Survey for Maps 8.1, and 9.3 to 9.12, Crown copyright reserved.
United States Geological Survey for Map 8.3.
McGraw-Hill International Book Company for Map 8.4.
Commission for the European Communities for Figure 8.3 and Map 8.2.
Keter Publishing House Jerusalem Ltd for Maps 9.14 and 9.15.

References

ACORD (Advisory Council on Research and Development for Fuel and Power) (1982). *Summary of Advice to the Secretary of State for Energy on his Research and Development Programme on Renewable Energy Sources.* Department of Energy, London.

Ader, G., Bridgewater, A. V., and Hatt, B. (1978). *Conversion of Biomass to Fuels by Thermal Processes.* Energy Technology Support Unit, Harwell, Oxfordshire.

Ader, G. and Wheatley, B. I. (1979). *Conversion of Biomass to Fuels by Anaerobic Digestion.* Energy Technology Support Unit, Harwell, Oxfordshire.

Alford, R. F. G. and editorial committee (1971). *The British Economy. Key Statistics 1900–1970.* Times Newspapers Ltd., London.

Allen, J. and Bird, R. A. (1977). *The Prospect for the Generation of Electricity from Wind Energy in the United Kingdom.* Energy Paper No. 21. H.M.S.O., London.

Bakke, P. and working party (1975). *Energy Conservation: A Study of Energy Consumption in Buildings and Possible Means of Saving Energy in Housing.* H.M.S.O., London.

Baldwin, J. L. (1968). *Climatic Atlas of the United States.* U.S. Department of Commerce, Washington, D.C., reprinted by the National Oceanic and Atmospheric Administration, 1979.

Bernshtein, L. B. (1961). *Tidal Energy for Electric Power Plants.* Gosudarstvennoe Energeticheskoe Izdatelstvo, Moskova–Leningrad. Translated by the Israel Program for Scientific Translations, 1975, Jerusalem.

Best, R. H. (1981). *Land Use and Living Space.* Methuen and Co. Ltd., London.

Bigelow, H. B. and Edmondson, W. T. (1947). *Wind Waves at Sea, Breakers and Surf.* Hydrographic Office Publication No. 602, United States Naval Oceanographic Office, Washington, D.C.

Bowditch, N. (1966). *American Practical Navigator.* Hydrographic Office Publication No. 9, United States Naval Oceanographic Office, Washington, D.C.

Bowen, M. G. (1978). *Canada's Forest Inventory 1976.* Canadian Forestry Service, Environment Canada, Ottowa.

Brown, G. C. and Mussett, A. E. (1981). *The Inaccessible Earth.* George Allen and Unwin, London.

Bureau of the Census (1981). *Statistical Abstract of the United States.* United States Department of Commerce, Washington, D.C.

Burley, A. J. and Edmunds, W. M. (1978). *Catalogue of Geothermal Data for the Land Area of the United Kingdom.* Institute of Geological Sciences, London.

Burt, W. V. (1954). *Albedo Over Wind Roughened Water.* Journal of Meteorology, Volume 11, Boston, Massachussetts.

Bush, R. P. and Chadwick, A. T. (1979). *A Disaggregated Model of Energy Consumption in UK Industry.* Energy Technology Support Unit, Harwell, Oxfordshire.

Canadian Normals (1975). *Wind.* Volume 3. Environment Canada, Toronto.

Central Statistical Office (1981a). *Regional Trends 1981.* H.M.S.O., London.

Central Statistical Office (1981b). *Economic Trends Annual Supplement.* H.M.S.O., London.

Centre for Agricultural Strategy (1980). *Strategy for the UK Forest Industry.* Report No. 6. Reading, England.

Chelton, D.B., Hussey, K.J., and Parke, M.E. (1981). Global satellite measurements of water vapour, wind speed and wave height. *Nature*, **294**, London.

CNCIHD (Canadian National Committee for the International Hydrological Decade) (1972). *Discharge of Selected Rivers of Canada*. Secretariat of CNCIHD, Ottowa.

Collingbourne, R.H. (1975). *United Kingdom Solar Radiation Network and the Availability of Solar Radiation Data from the Meteorological Office for Solar Energy Applications.* Proceedings of Conference on UK Meteorological Data and Solar Energy Applications, UK Section of the International Solar Energy Society, London.

CONAES (Committee on Nuclear and Alternative Energy Systems. Demand and Conservation Panel) (1979). *Alternative Energy Demand Figures to 2010*. National Research Council, National Academy of Sciences, Washington, D.C.

Cooper, J.P. (1975). *Photosynthesis and Productivity in Different Environments*. Cambridge University Press, London.

Cowley, J.P. (1978). The distribution over Great Britain of global solar radiation on a horizontal surface. *The Meteorological Magazine*, **107**, Number 1277, H.M.S.O., London.

Cowley, J.P. (1979). *Solar Radiation Measurements and Archives in the UK Meteorological Office, Bracknell.* Proceedings of Conference on Meteorology for Solar Energy Applications, UK Section of the International Solar Energy Society, London.

Dawson, J.K. (1979). *Wave Energy*. Energy Paper No. 42. H.M.S.O., London.

Deitrich, G. (1950). *Die naturlichen Regionen von Nord- und Ostsee auf hydrographischer Grundlage*. Keiler Meeresforschungen Band VII, Heft 2, Kiel.

Department of Agriculture and Fisheries for Scotland (1980). *Agricultural Statistics Scotland 1978*. H.M.S.O., Edinburgh.

Department of Energy (1973). *United Kingdom Energy Statistics*. H.M.S.O., London.

Department of Energy (1978). *Re-organisation of the Electricity Supply Industry in England and Wales*. Cmnd. 7134. H.M.S.O., London.

Department of Energy (1980). *Digest of United Kingdom Energy Statistics*. H.M.S.O., London.

Department of Energy (1981). *Digest of United Kingdom Energy Statistics*. H.M.S.O., London.

Doodson, A.T. and Warburg, H.D. (1941). *Admiralty Manual of Tides*. H.M.S.O., London.

Duffie, J.A. and Beckman, W.A. (1974). *Solar Energy Thermal Processes*. John Wiley and Sons Inc., New York.

Duncan, C.H., Willson, R.C., Kendall, J.M., Harrison, R.G., and Hickey, J.R. (1982). Latest rocket measurements of the solar constant. *Solar Energy*, **28**, Dublin.

Dunning, F.W. (1966). *Tectonic Map of Great Britain and Northern Ireland*. Ordnance Survey, Chessington, Surrey.

Edison Institute (1980). *Statistical Yearbook of the Electric Utility Industry – 1979*. Edison Electric Institute, Washington D.C.

Energy Policy Sector (1976). *An Energy Strategy for Canada: Policies for Self Reliance*. Department of Energy, Mines and resources, Ottowa.

Forest Service (1977). *The Nation's Renewable Resources. An Assessment. 1975*. United States Department of Agriculture, Washington, D.C.

Francis, G.H. (1980). *Use of Vegetable and Animal By-products*. Report of Symposium on By-products and Wastes in Animal Feeding. British Society of Animal Production. Occasional Publication No. 3. Reading, Berkshire.

Garland, G.D., and Lennox, D.H. (1962). Heat flow in Western Canada. *Geophysical Journal of the Royal Astronomical Society*, **6**, Oxford.

Garnish, J.D. (1976). *Geothermal Energy; The Case for Research in the United Kingdom*. H.M.S.O., London.

Geothermal Energy Project (1982). *Bimonthly Information Circular Number 8*. Camborne School of Mines, Camborne, Cornwall.

Gibrat, R. (1966). *L'Energie des Marees*. Presses Universitaires de France, Paris.

Godin, G. (1980). *Cotidal Charts for Canada*. Manuscript Report Series No. 55. Marine Sciences and Information Directorate, Department of Fisheries and Oceans, Ottowa.

Golding, E. W. (1976). *The Generation of Electricity by Wind Power*. E. and F. N. Spon Ltd., London.

Haenel, R. (1980). *Atlas of Sub-surface Temperatures in the European Community*. Directorate-General for Research Science and Education, Commission of the European Communities, Luxembourg.

Hatfield, J. L., Giorgis, R. B., and Flocchini, R. G. (1981). A simple solar radiation model for computing direct and diffuse spectral fluxes. *Solar Energy*, **27**, Dublin.

Hodges, D. and Horton, A. (1979). *Milton Keynes Solar House – Performance of Solar Heating System*. Built Environment Research Group of the Polytechnic of Central London in association with the Milton Keynes Development Corporation, London.

Hughes, R. (1977). *The Fate of Cereal Straw – 1976*. Report on Straw Utilisation Conference 1977. Ministry of Agriculture Fisheries and Food, London.

Hydrographic Department (1971). *Southern North Sea. Co-tidal and Co-range Chart*. Admiralty Chart No. 5059. Ministry of Defence, Taunton, Somerset.

Hydrographic Department (1974a). *Dungeness to Hoek van Holland. Co-tidal and Co-range Chart*. Admiralty Chart No. 5057. Ministry of Defence, Taunton.

Hydrographic Department (1974b). *The British Isles and Adjacent Waters. Co-tidal and Co-range Lines*. Admiralty Chart No. 5058. Ministry of Defence, Taunton.

Hydrographic Department (1976). *The British Isles*. Admiralty Chart No. 2. Ministry of Defence, Taunton.

Hydrographic Department (1979a). *European Waters*. Tide Tables, Volume 1. Ministry of Defence, Taunton.

Hydrographic Department (1979b). *The Atlantic and Indian Oceans*. Tide Tables, Volume 2. Ministry of Defence, Taunton.

Hydrographic Department (1979c). *The Pacific Ocean and Adjacent Seas*. Tide Tables, Volume 3. Ministry of Defence, Taunton.

Hyndman, R. D., Jessop, A. M., Judge, A. S., and Rankin, D. S. (1979). Heat flow in the Maritime Provinces of Canada. *Canadian Journal of Earth Sciences*, **16**, No. 6, Ottowa.

International Energy Agency (1977). *World Energy Outlook*. Organisation for Economic Co-operation and Development, Paris.

International Energy Agency (1981a). *Energy Balances of OECD Countries. 1975/79*. Organisation for Economic Co-operation and Development, Paris.

International Energy Agency (1981b). *Energy Statistics 1975/1979*. Organisation for Economic Co-operation and Development, Paris.

Iseri, K. T. and Langbein, W. B. (1974). *Large Rivers of the United States*. United States Geological Survey Circular 686, Department of the Interior, Washington, D.C.

Jerlov, N. G. (1976). *Marine Optics*. Elsevier Scientific Publishing Company, Amsterdam.

Jessop, A. M. and Lewis, T. (1978). Heat flow and heat generation in the Superior Provinces of the Canadian Shield. *Tectonophysics*, **50**, Amsterdam.

Judge, A. (1974). *Geothermal Measurements in Northern Canada*. Proceedings of the Symposium on the Geology of the Canadian Arctic. Geological Association of Canada and the Canadian Society of Petroleum Geologists, Toronto.

Justin, B., Guy, A. G., and Shaw, G. (1980). *Monitoring Solar Walls in Occupied Houses*. Research and Development Laboratories of Pilkington Brothers Ltd., Latham, Lancashire.

King, P.B. (1969). *Tectonic Map of North America*. North and South Sheets. United States Geological Survey, Arlington, Virginia.

Krummel, O. (1911). *Handbuch der Ozeanographie*. J. Engelhorns, Stuttgart.

Lacy, R.E. (1977). *Climate and Building in Britain*. H.M.S.O., London.

Lawrie, T. et al. (1944). *Development Scheme*. North of Scotland Hydro-electric Board, Edinburgh.

Leach, G., Lewis, C., Romig, F., van Buren, A., and Foley, G. (1979). *A Low Energy Strategy for the United Kingdom*. Science Reviews Ltd.; London.

Lewis, T.J., Judge, A.S., and Souther, J.G. (1979). Possible geothermal resources in the Coast Plutonic Complex of Southern British Columbia, Canada. *Pure and Applied Geophysics*, **117**, Basel.

Lindal, B. (1973). *Industrial and Other Applications of Geothermal Energy*. Geothermal Energy Review of Research and Development, Earth Sciences 12. U.N.E.S.C.O., Paris.

Long, G. (1976). *Solar Energy: Its Potential Contribution Within the United Kingdom*. Energy Paper No. 16. H.M.S.O., London.

McAuliff, C.A. (1980). *Hydrogen and Energy*. Macmillan Press Ltd., London.

Mangarella, P.A. and Heronemus, W.E. (1979). *Thermal Properties of the Florida Current as Related to Ocean Thermal Energy Conversion*. Solar Energy, **22**, Dublin.

Marcus, S.O. and collaborators (1973). *Environmental Conditions Within Specified Geographical Regions*. Environmental Data Service of the National Oceanic and Atmospheric Administration, United States Department of Commerce, Washington, D.C.

Maxson, W.P. and committee (1972). *Hydroelectric Power Resources of the United States. Developed and Undeveloped*. Ref. P-42. Federal Power Commission, Washington, D.C.

Meinel, A.B. and Meinel, M.P. (1977). *Applied Solar Energy*. Addison-Wesley Publishing Company Inc., Reading, Massachusetts.

Ministry of Agriculture, Fisheries and Food (1981a). *Agricultural Statistics 1978/1979*. H.M.S.O., London.

Ministry of Agriculture, Fisheries and Food (1981b). *Agricultural Statistics England 1978/1979*. H.M.S.O., London.

Muffler, L.J.P. (ed.) (1979). *Assessment of Geothermal Resources of the United States – 1978*. Geological Survey Circular 790. United States Geological Survey, Arlington, Virginia.

Muffler, P. and Cataldi, R. (1978). Methods of regional assessment of geothermal resources. *Geothermics*, **7**, Elmsford, New York.

Munn, L.C. and committee (1980). *Land Use in Canada. The Report of the Interdepartmental Task Force on Land Use Policy*. Lands Directorate, Environment Canada, Ottowa.

Naval Weather Service Detachment (1974). *North Atlantic Ocean*. United States Navy Marine Climatic Atlas of the World, Volume 1. Director, Naval Oceanography and Meteorology, Washington, D.C.

Naval Weather Service Detachment (1977). *North Pacific Ocean*. United States Navy Marine Climatic Atlas of the World, Volume 2. Director, Naval Oceanography and Meteorology, Washington, D.C.

NORPAC (1960). *Oceanic Observations of the Pacific; 1955, The Norpac Atlas*. University of California Press and University of Tokyo Press, Berkley and Tokyo.

Oceanographic Office (1976). *Oceanographic Atlas of the North Atlantic Ocean. Section II, Physical Properties*. Publication No. 700. United States Naval Oceanographic Office, Washington, D.C.

Ordnance Survey (1982). *The Ordnance Survey Atlas of Great Britain*. Country Life Books, London.

Oxburgh, E.R., Richardson, S.W., Bloomer, J.R., Martin, A., and Wright, S. (1977). *Sub-surface Temperatures from Heat Flow Studies in the United Kingdom*. Proceedings of Seminar on Geothermal Energy. Directorate-General for Research Science and Education, Commission of the European Communities, Luxembourg.

Paltz, W. and Chartier, P. (1980). *Energy from Biomass in Europe*. Applied Science Publishers Ltd., London.

Parker, S. P. and editorial staff (1982). *McGraw-Hill Encyclopedia of Science and Technology*. McGraw-Hill Inc., New York.

Pierson, W. J., Neuman, G., and James, R. W. (1955). *Observing and Forecasting Ocean Waves*. Hydrographic Office Publication No. 603. United States Naval Oceanographic Office, Washington, D.C.

Roberts, L. (1979). *Radioactive Waste – Policy and Perspective*. Atom, No. 267, London.

Robinson, M. K. (1976). *Atlas of North Pacific Ocean Monthly Mean Temperatures and Mean Salinities of the Surface Layer*. NOO RP–2. Naval Oceanographic Office, Bay St Louis, Mississippi.

Robinson, M. K., Bauer, R. A., and Schroeder, E. H. (1979). *Atlas of North Atlantic – Indian Ocean Monthly Mean Temperatures and Mean Salinities of the Surface Layer*. NOO RP–18. Naval Oceanographic Office, Bay St Louis, Mississippi.

Rutherford, I. (1976). *Straw Yield Investigations 1975*. Report on Straw Utilisation Conference 1976. Ministry of Agriculture Fisheries and Food, London.

Salter, S. H. (1974). Wave power. *Nature*, **249**, London.

Sass, J. H., Blackwell, D. D., Chapman, D. S., Costain, J. K., Deker, E. R., Lawver, L. A., and Swanberg, C. A. (1981). Heat flow from the crust of the United States. In: *Physical Properties of Rocks and Minerals*, Touloukian, Y. S., Judd, W. R., and Roy, R. F. (Eds.), McGraw-Hill Book Company, New York.

Saunders, P. A. H. (1981). *The Management of High Level Waste and Its Environmental Impact*. Proceeding of the Conference on the Environmental Impact of Nuclear Power, British Nuclear Energy Society, London.

Schlesinger, B. and committee (1980). *Gas Facts 1980*. American Gas Association, Arlington, Virginia.

Secretary of State for Energy (1978). *Energy Policy. A Consultative Document*. Cmnd. 7101. H.M.S.O., London.

Select Committee on Science and Technology (1977a). *The Development of Alternative Sources of Energy for the United Kingdom. Memorandum and Supplementary Memorandum submitted by Engineering and Power Development Consultants Ltd*. Third Report from the Select Committee on Science and Technology. H.M.S.O., London.

Select Committee on Science and Technology (1977b). *The Exploitation of Tidal Power in the Severn Estuary*. Fourth Report of the Select Committee on Science and Technology. H.M.S.O., London.

Severn Barrage Committee (1981). *Tidal Power from the Severn Estuary*. H.M.S.O., London.

Shellard, H. C. (1968). *Tables of Surface Wind Speed and Direction over the United Kingdom*. Report 792, H.M.S.O., London.

Slesser, M. and Lewis, C. (1979). *Biological Energy Resources*. E. and F. N. Spon Ltd., London

Statistics Canada (1979a). *Estimate of Principal Field Crops 1979*. Catalogue 22–002. Ministry of Trade and Commerce, Ottawa.

Statistics Canada (1979b). *Gas Utilities. December 1979*. Catalogue 55–002. Treasury Board, Ottawa.

Statistics Canada (1980). *Livestock and Animal Products Statistics*. Catalogue 23–203. Ministry of Supply and Services, Ottawa.

Sutcliffe, J. V. and team (1975). *Flood Studies Report*. Natural Environment Research Council, London.

Sverdrup, H. U., Johnson, M. W., and Fleming, R. H. (1942). *The Oceans*. Prentice Hall Inc., Englewood Cliffs, New Jersey.

Tagg, J. R. (1957). *Wind Data Related to the Generation of Electricity by Wind Power*. Technical Report C/T 115. The British Electrical and Allied Industries Research Association, London.

Taylor, B. J., Burgess, I. C., Lord, D. H., Mills, D. A. C., Smith, D. B., and Warren, P. T. (1971). *British Regional Geology. Northern England*. H.M.S.O., London.

Thekaekara, M. P. and Drummond, A. J. (1971). Standard values for the solar constant and its spectral components. *Nature Physical Science*, **229**, London.

Titus, R. L. and Truhlar, E. J. (1969). *A New Estimate of Average Global Solar Radiation in Canada*. Meteorological Branch, Department of Transport, Toronto.

Tung, T. (1983). Personal communication.

United States Department of Agriculture (1981). *Agricultural Statistics 1981*. United States Department of Agriculture, Washington, D.C.

Water Data Unit (1978). *Surface Water: United Kingdom 1971–73*. Department of the Environment, London.

Welsh Office (1980). *Welsh Agricultural Statistics 1980*. H.M.S.O., Cardiff.

Wilson, E. M., Haine, C. S., and Hamer, R. H. (1980). *Small Scale Hydroelectric Potential of Wales*. Chadwick-Healey Ltd., Cambridge.

Winter, A. J. B. (1980). The UK wave energy resource. *Nature*, **287**, London.

Bibliography

Considene, D.M., Crawford, H.B., Eisler, W., and Shaw, W. (1977). *Energy Technology Handbook.* McGraw-Hill Inc., New York.

Crabbe, D. and McBride, R. (1978). *The World Energy Book.* Kegan Paul Ltd., London.

Foley, G. and Nassim, C. (1976). *The Energy Question.* Penguin Books Ltd., Harmondsworth.

Lapedes, D.N. and editorial staff (1976). *Encyclopedia of Energy.* McGraw-Hill Inc., New York.

Odum, T.O. (1971). *Environment, Power and Society.* John Wiley and Sons Inc., New York.

Patterson, W.C. (1976). *Nuclear Power.* Penguin Books Ltd., Harmondsworth.

Sorensen, B. (1979). *Renewable Energy.* Academic Press Inc. Ltd., London.

Index

Adirondak Mountains, 26
aerogenerator
   Aldeborough, 51
   General Electric Company, 51, 53
Afsluitdijk, 151
agricultural fertilizers, 12
air mass, definition of, 17
Akkadians, 15
Alabama, 52, 140
Alaska, 3, 25, 26, 27, 52, 53, 90, 91, 110, 139, 151
Alberta, 26
Aleutian Islands, 2, 151
anaerobic digestion, 97, 101, 102, 103, 106
Anchorage, 151, 161
Annapolis Bay, 152, 161
Appalachian Mountains, 3, 23, 52, 90
Apollo, 15
aquifers, 130, 135, 136, 137, 138, 140
Ardnamurchan, 137
Arizona, 3
Arran, 137
asthenosphere, 129
atmospheric
   aerosols, 17
   dust, 9, 17
   jet streams, 3
   tides, 146

Baffin Bay, 52, 151
Baffin Island, 52, 53, 90
   Lake Harbour, 152
barrages
   river, 87, 165
   tidal, 1, 14, 148, 149, 150, 151
basalt, 130
basic oxygen steel manufacture, 163
Bay of Fundy, 3, 152, 161
Beaufort Sea, 26
Bebington, 25
Ben Nevis, 2
Bering Strait, 151
Bermuda Islands, 80
Bimini, 83
biofuels, 3, 9, 95–128, 165, 168
   feedstock, 97, 109
      barley, 99, 100, 101, 109
      Brussels sprouts, 101
      catch crops, 101, 104, 106

      cereal straw, 97, 100, 101, 103, 104, 110
      forestry wastes, 97, 99, 101, 103, 104, 106, 109
      kale, 103
      livestock manure, 97, 104, 106, 109, 110
      maize, 109
      oats, 99–101
      peas, 101
      potatoes, 97, 101
      rape, 103
      sugar beet, 97, 101
      tillage crops, 97, 101, 102, 104, 106
      wheat, 99, 100, 101, 105
biological hazard, 10, 12
   actual, 12
   potential, 12
biomass, 95
biosphere, 10, 13
boreholes, 128, 130, 132, 135
   Baden No. 1, 130
   Rosemanowes, 131
   Seal Sands, 130
   Winterborne Kingston, 131
British Columbia, 3
Brooks Range, 139
buildings
   hot water supply, 24
   space heating, 24, 132, 134, 163, 165, 168
   useful life of, 164
Burntcoat Head, 152
Buxton, 138

California, 110
calms, 48
Canadian arctic archipelago, 2, 25, 26, 52, 53, 139, 151
Canadian Shield, 3
cancer, 12
Cape Cod, 152
Cape Columbia, 25
Cape Hatteras, 80
carbohydrates, 95
carbon dioxide, 9, 95
Caribbean Sea, 26, 80
Carmarthen Bay, 150, 157
Carnmenellis, 137
Cascade Mountains, 25
Central America, 80
Central Electricity Generating Board, 9, 167

197

Channel Islands, 23, 150
Chignecto Bay, 152, 161
chlorophyll, 95
chloroplasts, 95
clay, 130
coal, 2, 7, 9, 12, 13, 14, 168, 170, 172
Cobscook Bay, 152
coke, 9
Colorado, 25
combustion, 95, 97
coriolis force, 147
Cornish granite, 137
Cornwallis Island, 139
Costa Hill, 50
cropland, 96
Cumbria, 137

Dakota, North, 140
Davis Strait, 151
Death Valley, 3
Derbyshire, 23
desert, 3
Devon, 137
Durham, 137

Earth
  crust, 7, 130, 131
  inner core, 129, 130
  lithosphere, 129, 130
  magnetic field, 129
  mantle, 129, 130
  outer core, 129
  structure, 129
  surface heat flow, 130, 131, 137, 138, 140
  tectosphere, 129
  tides, 146
Edmonton, 26
El Paso, 25
electric road vehicles, 167
electricity, 9, 13, 90, 91, 170
  demand for, 7, 9
  generation of, 9, 163, 165, 172
  storage of, 167
  tidal, 149–152, 171, 172
Electricity Acts, 167, 168
Electricity Corporation, 168
Ellesmere Island, 25, 26
energy
  capital endowment of, 2, 7
  conservation, 97, 163, 164
  conversion processes, 3–5, 79
  crops, 96
  definition of, 1
  delivered, 7
  demand, 2, 7
  income of, 7
  primary, 7, 9, 165, 167, 168
    definition of, 9
  renewable, definition of, 2
  secondary, definition of, 9
  waste, 7, 9, 165, 167, 168
energy cost of economic activity, 163, 164
energy flow
  North American economy, 8, 169
  United Kingdom economy, 6, 166
Energy Technology Support Unit, 69
English Channel, 50, 69, 79, 80
English Lake District, 23, 137

Fire Island, 151
fish farming, 24
Fleetwood, 51
Florida, 3, 52, 79, 80
Florida current, 83
Florida Strait, 81
food chain, human, 12
forests, 2, 3, 48, 96, 97
  Belgium, 98
  France, 98
  Germany, 98
  Irish Republic, 98
  North America, 109, 111
  United Kingdom, 96, 98, 99
fossil fuels, 2, 5, 7, 13, 165, 170
  definition of, 9
Frobisher Bay, 52
fuel
  consumption of, 163, 165, 168
  demand for, 164, 165, 168
  price of, 170, 171

gales, 48
Galloway, 137
gas
  biogas, 97, 98, 101, 104, 106, 110, 167
  manufactured, 9, 170
  natural, 2, 7, 9, 13, 97, 98, 106, 167, 168, 170, 172
gas plant
  biogas, 1, 3
  hydrogen, 152, 165, 167
gasification, 97
  oxygen, 99, 101
General Strike, 163
Georgia, 52, 140
geotectonic activity, 2, 139
geothermal energy, 3, 9, 129–144, 165
geothermal field
  Geysers, 138
  Lassen, 138
  Mud Volcano, 138
  Yellowstone, 138
glass, 21, 24
  transmittance of, 22
  water white, 21, 22
glucose, 95

Gower Peninsula, 51
gradient
   geothermal
      hyperthermal, 132, 138
      normal, 132, 134
      semi-thermal, 132, 134
   geothermal temperature, 131, 132–137, 138
   ocean salinity, 3
   ocean thermal, 2, 3, 78, 79, 80, 82
granite, 11, 12, 130, 131, 134, 137
grassland, 3, 96, 104, 106
gravitation, force of, 2, 145–147
Great Plains
   Canada, 26
   United States, 25, 52
Great War, 163
greenhouse effect, 21, 24
Greenland, 26
gross domestic product, 163, 164
groundwater, 12, 131, 137, 138
Gulf of Maine, 152
Gulf of Mexico, 25, 79, 80, 83, 152

Hampshire, 138
Hawaii, 139
heat engines, 79
Hebrides, 50, 52, 68, 69, 70, 80, 137, 138
helium, 15
Hensbarrow, 137
horticulture, 24
Hudson Bay, 26, 52
Hudson Strait, 152
hydrofracturing, 134
hydrogen, 15, 152, 165, 167
hydrological cycle, 87

Idaho, 25
Ijsselmeer, 151
Illinois, 110
International Commission on Radiological
   Protection, 11
International Institute of Environment and
   Development, 163
iodine–129, 10
iodine–131, 10
Irish Sea, 80
iron, 129

Jersey, 23
Joule, James, 1

Keewatin, 52, 53
Kentucky, 140
Kew, 20
krypton–85, 10

La Grande Riviere, 91
Labrador, 52, 71

Lake Winnipeg, 26
land use
   North America, 109
   United Kingdom, 96
Lands End, 137
lignite, 9
limestone, 130, 132, 135
lithosphere, 129, 130
livestock
   cattle, 104
   chickens, 104, 106
   pigs, 104, 106
   populations, 105, 110
   sheep, 104
Lochaber, 137
London, 17, 18, 23, 52, 150, 151
Los Alamos Scientific Laboratory, 134
Louisiana, 25, 110

magma, 137
Maine, 26
Matlock, 138
Menai Strait, 150, 156
metal ore, smelting of, 152
Meteorological Office, 20, 48, 69
methane, 97
Mexico, 2, 25
Miami, 83
Milton Keynes, 25
Minas Basin, 152, 161
Mississippi, 52
Mohorovicic discontinuity, 130
Montana, 25
moon, 145, 146
moorland, 96, 109
Morecambe Bay, 137, 150, 156
Mount McKinley, 3
mudstone, 130
Mull, 137
Mynedd Analog, 50

National Coal Board, 9
National Radiation Centre, 20
Netherlands, 151
New Mexico, 25
New York State, 26
Newfoundland, 2, 52, 71
nickel, 129
nitrogen, 95
North Atlantic Drift current, 2
North Atlantic Ocean, 2, 3, 26, 69, 70
North Pacific Ocean, 3, 25, 71
North Sea, 13, 69, 71, 80, 97
Northern Ireland, 90, 99, 100, 104, 106, 110,
   137, 150, 167
Northumberland, 137
Norway, 9
Nottinghamshire, 137

Nova Scotia, 52, 152
nuclear fuel, 10, 11
   cycle, 10–12
nuclear power, 9, 10, 12, 13, 169, 170

ocean currents, 47, 79
ocean temperatures, seasonal variation of, 81, 82
ocean thermal conversion machines, 79, 81
   deployment of, 83
ocean thermal energy, 3, 77–85, 165
ocean weather ship *India*, 68, 70
Ohio, 110
oil, 2, 7, 9, 12, 170, 172
open hearth furnace, 163
Organization of Petroleum Exporting Countries, 170
Owers, Bank, 50

Passamaquoddy Bay, 152, 161
Pennine Mountains, 23, 48, 50
Pennsylvania, 140
petrochemical industry, 163
petroleum, 13
phosphorous, 95
photosphere, 16
photosynthesis, 95
plate tectonics, 129
plutonium, 9, 10, 13
plutonium–239, 10
pollution, 9, 23, 95
   thermal, 10
poly-aromatic hydrocarbons, 12
Poole, 138
potassium, 130
power, definition of, 1
power plant
   combined heat and power, 163, 167, 168
   efficiency of, 9, 167
   electric, 130, 167
   geothermal, 1, 7, 138, 139
   nuclear, 10
   nuclear AGR, 11
   thermal, 9, 12, 167
   tidal, 149
   tidal, merit order of, 151
   waves, 70
Prairies, United States, 25
Puget Sound, 52
pyrolysis, 97

Queen Charlotte Island, 52

Ra, 15
radioactive half life, definition of, 10
radioactive wastes, 10, 11
   glassification of, 11
   underground repositories for, 11, 12
radioactivity, 2, 10, 130, 137
   release of, 13
radiotoxins, 11
   maximum permissible concentrations of, 11
reservoirs
   river, 87
   tidal, 148, 149, 150
Rhossili Down, 50, 51
Rhum, 137
risk analysis, 12
River
   Churchill, 91
   Clyde, 89
   Columbia, 91
   Dee, 150, 157
   Humber, 150, 159
   Mackenzie, 139
   Mersey, 150, 157
   Mississippi, 25, 81
   Mississippi–Missouri, 90
   Rance, 148, 149, 150, 151
   St Lawrence, 91
   Severn, 2
   Tweed, 89
river catchment area, 88
river energy, 3, 7, 9, 87–94, 165
rivers
   gross energy of, 87, 89, 92
   largest, 88, 91
   most energetic, 89, 92
road transport, 163
rock formations
   igneous, 130, 134, 137
   porosity of, 134, 135, 136
   sedimentary, 130, 134
   sedimentary basins, 130, 131, 135
      Alaska Coastal, 139
      Alberta, 140
      Allegheny, 140
      Anadarko, 140
      Black Warrior, 140
      Cheshire, 137, 138
      Lincolnshire, 137
      Michegan, 140
      North Gulf of Mexico, 15, 130, 131, 137, 140
      Peel, 139
      Wessex, 131, 137, 138
      Williston, 140
      Worcestershire, 138
Rocky Mountains, 2, 3, 25, 52, 90, 109, 139
rough grazing land, 96
Royal Sovereign Bank, 50

Sacramento Valley, 52
St Bees Head, 137
San Francisco, 52, 138
sandstone, 130, 132, 135
Saskatchewan, 26

Scilly Isles, 70, 137
Scotland, 7, 50, 90, 99, 104, 106, 150, 167
    Highlands of, 23, 48, 50, 89
    Lowlands of, 23
    Southern Uplands of, 23, 48, 50
Seasat mission, 70
seawater, transparency of, 77, 78
Second World War, 163
Sequoia Gigantea, 95
Sequoia National Park, 95
Severn Estuary, 147, 149, 150, 151, 158, 170
Shamash, 15
Shetland Isles, 23
Sierra Nevada, 25
Skye, 137
solar constant, 16, 18
solar energy, 3, 15–46, 165, 168
   active collection of, 24, 25
   collection devices, 1, 21, 24
   collection devices, selective surfaces, 22, 23, 24
   passive collection of, 24, 25
solar flux
   extra terrestrial, 3, 16
   terrestrial, 17, 24, 70
solar radiation
   absorbtion of, 21, 77
   atmospheric transmission of, 18
   comparison of insolation levels, 23, 25–27
   data, 20, 21
   diffuse, 18, 21, 23, 24
   direct, 18, 21, 24
   diurnal variation of, 19
   global, 18, 21, 23
   recording stations, United Kingdom, 28
   seasonal variation of, 19, 20
   spectrum of, 16, 17
Solway Firth, 150, 155
South Uist, 70
Staffordshire, 23
standard atmosphere, 47
standard of living, 163
Strangford Lough, 150, 151, 155
strontium–90, 10
sulphur, 9, 129
Sumerians, 15
sun, 2, 15, 145
   core, 15, 16
   description of, 15
   disc, 15
   energy production of, 16
   fusion reaction, 15
   light, 95
'sunbowl', 25
sunshine
   duration data, 21
   recording stations, United Kingdom, 29
Sussex, West, 138

swamp, 3
Sweden, 9
swimming pool heating, 24

Teesside, 130
temperatures, industrial processes, 4, 5, 132
Tennessee, 140
Texas, 2, 25, 52, 130, 137, 140
Thames Estuary, 50, 150, 151, 159
thermodynamics, second law of, 1
thorium, 130
tidal
   energy, 3, 95, 145–161, 165
   range, 2, 3, 148, 150, 151, 152
   streams, 79, 80, 146, 147
   tractive force, 146
tides
   atmospheric, 146
   earth, 146
   neap, 147
   raising force, 145–146
   resonance of, 147
   semi-diurnal, 147
   spring, 147
toxic potential
   chemical, 12
   definition of, 11
   index of, 11
   radioactive, 12
tundra, 3
turbines
   gas, 5
   steam, 5
   tidal, 1, 147, 150

ultraviolet light, 12
Ungava Bay, 152
United States Geological Survey, 138
uranium, 9, 10, 12, 130
   ore, 11
uranium–235, 10
Utah, 25
Utu, 15

Virginia, 52
   West, 140
volatile solids, 97
vulcanism, 2, 130

Wales, 50, 90, 99, 100, 104, 106, 138
Wash, The, 150, 158
waste land, 96, 109
water, electrolytic dissociation of, 152, 165
Watt, James, 1
wave energy, 3, 65–76, 165
   deployment of machines, 70, 71
wave height, 3
   significant, 68, 70

Waverider, 69
waves, 2, 47, 65
    data collection, 69
    description of, 65, 66
    energy content of, 66–68
    fetch, 67
    height of, 67, 68
    modelling of, 69
    speed of, 67
Weardale granite, 137, 138
Welsh Mountains, 23, 48
wind, 2, 47

wind damage, 48
wind energy, 3, 47–63, 165
    annual variation of, 49
    of a site, 50
wind speed
    distribution graphs, 50, 51, 54–58
    geographical comparison of, 52, 53
    variation with height, 48, 49
wind turbulence, 50
windmills, 1, 3, 50
work, definition of, 1
Wyoming, 25